ARTIFICIAL INTELLIGENCE IN COMPUTATIONAL ENGINEERING

ELLIS HORWOOD SERIES IN MECHANICAL ENGINEERING

Series Editor: J. M. ALEXANDER, formerly Stocker Visiting Professor of Engineering and Technology, Ohio University, Athens, USA, and Professor of Applied Mechanics, Imperial College of Science and Technology, University of London

The series has two objectives: of satisfying the requirements of postgraduate and mid-career engineers, and of providing clear and modern texts for more basic undergraduate topics. It is also the intention to include English translations of outstanding texts from other languages, introducing works of international merit. Ideas for enlarging the series are always welcomed.

Alexander, J.M.	**Strength of Materials: Vol. 1: Fundamentals; Vol. 2: Applications**
Alexander, J.M., Brewer, R.C. & Rowe, G.	**Manufacturing Technology Volume 1: Engineering Materials**
Alexander, J.M., Brewer, R.C. & Rowe, G.	**Manufacturing Technology Volume 2: Engineering Processes**
Atkins A.G. & Mai, Y.W.	**Elastic and Plastic Fracture**
Beards, C.F.	**Vibration Analysis and Control System Dynamics**
Beards, C.F.	**Structural Vibration Analysis**
Beards, C.F.	**Noise Control**
Beards, C.F.	**Vibrations and Control Systems**
Besant, C.B. & Lui, C.W.K.	**Computer-aided Design and Manufacture, 3rd Edition**
Borkowski, J. and Szymanski, A.	**Technology of Abrasives and Abrasive Tools**
Borkowski, J. and Szymanski, A.	**Uses of Abrasives and Abrasive Tools**
Brook, R. and Howard, I.C.	**Introductory Fracture Mechanics**
Cameron, A.	**Basic Lubrication Theory, 3rd Edition**
Collar, A.R. & Simpson, A.	**Matrices and Engineering Dynamics**
Cookson, R.A. & El-Zafrany, A.	**Finite Element Techniques for Engineering Analysis**
Cookson, R.A. & El-Zafrany, A.	**Techniques of the Boundary Element Method**
Ding, Q.L. & Davies, B.J.	**Surface Engineering Geometry for Computer-aided Design and Manufacture**
Edmunds, H.G.	**Mechanical Foundations of Engineering Science**
Fenner, D.N.	**Engineering Stress Analysis**
Fenner, R.T.	**Engineering Elasticity**
Ford, Sir Hugh, FRS, & Alexander, J.M.	**Advanced Mechanics of Materials, 2nd Edition**
Gallagher, C.C. & Knight, W.A.	**Group Technology Production Methods in Manufacture**
Gohar, R.	**Elastohydrodynamics**
Gosman, B.E., Launder, A.D. & Reece, G.	**Computer-aided Engineering: Heat Transfer and Fluid Flow**
Gunasekera, J.S.	**CAD/CAM of Dies**
Haddad, S.D. & Watson, N.	**Principles and Performance in Diesel Engineering**
Haddad, S.D. & Watson, N.	**Design and Applications in Diesel Engineering**
Haddad, S.D.	**Advanced Diesel Engineering and Operation**
Hunt, S.E.	**Nuclear Physics for Engineers and Scientists**
Irons, B.M. & Ahmad, S.	**Techniques of Finite Elements**
Irons, B.M. & Shrive, N.G.	**Finite Element Primer**
Johnson, W. & Mellor, P.B.	**Engineering Plasticity**
Kleiber, M.	**Incremental Finite Element Modelling in Non-linear Solid Mechanics**
Kleiber, M. & Breitkopf, P.	**Finite Element Methods in Structural Engineering: Turbo Pascal Programs for Microcomputers**
Leech, D.J. & Turner, B.T.	**Engineering Design for Profit**
Lewins, J.D.	**Engineering Thermodynamics**
Malkin, S.	**Materials Grinding: Theory and Applications of Mechining with Abrasives**
Maltbaek, J.C.	**Dynamics in Engineering**
McCloy, D. & Martin, H.R.	**Control of Fluid Power: Analysis and Design, 2nd (Revised) Edition**
Osyczka, A.	**Multicriterion Optimisation in Engineering**
Oxley, P.L.B.	**The Mechanics of Machining**
Piszcek, K. and Niziol, J.	**Random Vibration of Mechanical Systems**
Polanski, S.	**Bulk Containers: Design and Engineering of Surfaces and Shapes**
Prentis, J.M.	**Dynamics of Mechanical Systems, 2nd Edition**
Renton, J.D.	**Applied Elasticity**
Richards, T.H.	**Energy Methods in Vibration Analysis**
Ross, C.T.F.	**Computational Methods in Structural and Continuum Mechanics**
Ross, C.T.F.	**Finite Element Programs for Axisymmetric Problems in Engineering**
Ross, C.T.F.	**Finite Element Methods in Structural Mechanics**
Ross, C.T.F.	**Applied Stress Analysis**
Ross, C.T.F.	**Advanced Applied Stress Analysis**
Roy, D. N.	**Applied Fluid Mechanics**
Roznowski, T.	**Moving Heat Sources in Thermoelasticity**
Sawczuk, A.	**Mechanics and Plasticity of Structures**
Sherwin, K.	**Engineering Design for Performance**
Stupnicki, J.	**Stress Measurement by Photoelastic Coating**
Szczepinski, W. & Szlagowski, J.	**Plastic Design of Complex Shape Structured Elements**
Thring, M.W.	**Robots and Telechirs**
Walshaw, A.C.	**Mechanical Vibrations with Applications**
Williams, J.G.	**Fracture Mechanics of Polymers**
Williams, J.G.	**Stress Analysis of Polymers 2nd (Revised) Edition**

ARTIFICIAL INTELLIGENCE IN COMPUTATIONAL ENGINEERING

Editor

M. KLEIBER
Professor of Applied and Computational Mechanics
Institute of Fundamental Technological Research
Polish Academy of Sciences, Warsaw

ELLIS HORWOOD
NEW YORK LONDON TORONTO SYDNEY TOKYO SINGAPORE

First published in 1990 by
ELLIS HORWOOD LIMITED
Market Cross House, Cooper Street,
Chichester, West Sussex, PO19 1EB, England
A division of
Simon & Schuster International Group

© Ellis Horwood Limited, 1990

All rights reserved. No part of this publication may be
reproduced, stored in a retrieval system, or transmitted,
in any form, or by any means, electronic, mechanical,
photocopying, recording or otherwise, without the prior
permission in writing, from the publisher

Typeset in Times by Ellis Horwood Limited
Printed and bound in Great Britain
by Hartnolls, Bodmin

British Library Cataloguing in Publication Data

Artificial intelligence in computational engineering.
1. Engineering. Applications of artificial intelligence
I. Kleiber, M.
620'.0028'563
ISBN 0–13–048273–0

Library of Congress Cataloging-in-Publication Data

Artificial intelligence in computational engineering /
editor, M. Kleiber.
p. cm.
ISBN 0–13–048273–0
1. Engineering mathematics — Data processing.
2. Artificial intelligence.   I. Kleiber, Michal
TA331.A77   1990
620',001'51–dc20   89–71701
                                                    CIP

Contents

List of contributors 7

Preface 9

PART I — ASSESSMENT OF STRUCTURAL BEHAVIOUR 13

1.   OSEIS: A system for deriving qualitative seismic behaviour of structural systems from structural descriptions 15
   J. Ganguly, E. Kausel, D. Sriram (Massachusetts Institute of Technology)

2.   Knowledge-based systems in engineering risk control 37
   D. I. Blockley (University of Bristol)

3.   Structural assessment in a combined symbolic–numeric environment 59
   T. Krauthammer (University of Minnesota)

4.   A blackboard consultation system for constitutive modelling in solid mechanics 75
   J. R. Ambroziak, M. Kleiber (Polish Academy of Sciences)

5.   Interactive control of non-linear finite element calculations by an expert system 97
   P. Wriggers (Universität Hannover), N. Tarnow (University of California at Berkeley)

6.   Concepts for the application of AI techniques in computational mechanics 121
   D. Hartmann (Ruhr Universität Bochum)

PART II — DESIGN OPTIMIZATION 133

7.   On a knowledge-based user interface for the structural optimization system LAGRANGE 135
   K. Schittkowski (Universität Bayreuth), R. Zotemantel (Messerschmit–Bölkow–Blohm GmbH)

8.   Statistical machine learning for the cognitive selection of non-linear programming algorithms in engineering design optimization 147
   D. A. Hoeltzel, W., H. Chieng (Columbia University)

9 **A knowledge-based expert system for selection of slab type for multistory buildings** 159
   M. L. Das (University of Lowell), S. K. Ghosh (Portland Cement Association)

10 **Consultative expert systems for finite element based analysis and design of structure systems** 181
   P. Hajela (University of Florida), J. L. Chen (National Cheng Kung University)

11 **A knowledge-based framework for constraint activity identification in optimal design of aircraft structures** 209
   P. Y. Papalambros (The University of Michigan)

12 **Using artificial intelligence in an open software architecture for modelling in engineering** 227
   O. Aunay, S. Aunay, D. Chorlay, G. Touzot, M. Vayssade (Université de Technologie de Compiègne)

Index 253

List of contributors

Jacek R. Ambroziak
Institute of Fundamental Technological Research, Polish Academy of Sciences, Świętokrzyska 21, PL-00049 Warsaw, Poland

Olivier Aunay
Université de Technologie de Compiègne, Modèles Numériques en Mécanique, BP 649, 60206 Compiègne Cédex, France

Sylvain Aunay
Université de Technologie de Compiègne, Modèles Numériques en Mécanique, BP 649, 60206 Compiègne Cédex, France

D. I. Blockley
Department of Civil Engineering, University of Bristol, Bristol BS8 1TR, UK

Jahau Lewis Chen
Department of Mechanical Engineering, National Cheng Kung University, Tainen, Taiwan

W. H. Chieng
Department of Mechanical Engineering, Columbia University, New York, NY 10027, USA

Dominique Chorlay
Université de Technologie de Compiègne, Modèles Numériques en Mécanique, BP 649, 60206 Compiègne Cédex, France

Mukti L. Das
Department of Civil Engineering, University of Lowell, Lowell, MA 01854, USA

Jaideep Ganguly
Department of Civil Engineering, Massachusetts Institute of Technology, Cambridge, MA 02139, USA

Satyen K. Ghosh
Engineered Structures, Portland Cement Association, Skokie, Il., USA

Prabhat Hajela
Department of Aerospace Engineering, Mechanics and Engineering Science, University of Florida, Gainesville, FL 3261, USA

D. Hartmann
Fakultät für Bauingenieurwesen, Ruhr Universität Bochum, 4630 Bochum 1, FRG

D. A. Hoeltzel
Department of Mechanical Engineering, Columbia University, New York, NY 10027, USA

Eduardo Kausel
Department of Civil Engineering, Massachusetts Institute of Technology, Cambridge, MA 02139, USA

Michał Kleiber
Institute of Fundamental Technological Research, Polish Academy of Sciences, Świętokrzyska 21, PL-00049 Warsaw, Poland

Theodor Krauthammer
Department of Civil and Mineral Engineering, University of Minnesota, Minneapolis, MN 55455, USA

Panos Y. Papalambros
Department of Mechanical Engineering and Applied Mechanics, The University of Michigan, Ann Arbor, MI 48109-2125, USA

K. Schittkowski
Mathematisches Institut, Universität Bayreuth, 8580 Bayreuth, FRG

Duvvuru Sriram
Department of Civil Engineering, Massachusetts Institute of Technology, Cambridge, MA 02139, USA

N. Tarnow
Department of Civil Engineering, University of California at Berkeley, Berkeley, CA 94720, USA

Gilbert Touzot
Université de Technologie de Compiègne, Modèles Numériques en Mécanique, BP 649, 60206 Compiègne Cédex, France

Michel Vayssade
Université de Technologie de Compiègne, Modèles Numériques en Mécanique, BP 649, 60206 Compiègne Cédex, France

P. Wriggers
Institut für Baumechanik und Numerische Mechanik, Universität Hannover, 3000 Hannover, FRG

R. Zotemantel
Messerschmidt–Bölkow–Blohm GmbH, Aircraft Division, 8012 Ottobrunn, FRG

Preface

Despite all the reservations about what is loosely called artificial intelligence (AI) which are still being expressed in the community of researchers and engineers working in the field of computational engineering, there can be little doubt that AI will have in the future a major impact on the engineering profession. Although the computer-based technologies of the 1970s have provided effective tools for the development of useful large-scale programs in almost every field of engineeering, these technologies are clearly not adequate to deal with all such problems and there is enough evidence that at least some of them may be treated more effectively by completely new computer methods. From robotics to risk control and process diagnosis, from natural language processing to structural design optimization, few engineering disciplines seem to be able ultimately to escape the possibilities offered by the AI-based computing systems of the future.

Methods of computational engineering have been traditionally based on formalisms and laws of quantitative knowledge, which make it possible to formulate problem-governing equations and to work out analytical or numerical solution techniques for them. However, apart from the science of quantitative knowledge, it has been widely recognized that the art of qualitative knowledge is indispensible for solving large compound problems in modern engineering. Qualitative knowledge, as a rule employed only in the 'background' and thus virtually inaccessible to the casual observer, is decisive in splitting the overall problem into subtasks and in handling problems that are to difficult to describe in the precise language of formal theories. Typical problems we have in mind are those without purely algorithmic solutions such as preliminary and detailed design, large optimization problems, controlling parameters in large-scale non-linear computations and risk assessment.

So far, no monograph on the use of AI concepts in engineering has been published and the available publications are casually scattered over journals, conference proceedings and internal reports. By providing a collection of very representative articles this volume has been prepared to accelerate the closing of the communication gap between scientists working in different engineering areas and those using AI concepts. Particular emphasis is placed on subjects which traditionally rely on novel computer techniques such as computational mechanics, computer-aided design and optimization methods. The chapters included address successful applications of AI techniques to solve realistic engineering problems.

Because of the novelty of the subject, its enormous scope and the lack of broadly accepted methodologies, it was difficult to come up with a division of the volume into parts grouping contributions dealing with separate, clear-cut topics. The most logical structure of the book was finally believed to be achieved by splitting it into Part I, which groups contributions dealing with problems of engineering analysis, and Part II consisting of chapters devoted to design optimization.

The book opens with Chapter 1, written by J. Ganguly, E. Kausel and D. Sriram. They describe a knowledge-based system that can provide expert advice on the seismic behaviour of structural systems. The nature of the knowledge structures and the reasoning mechanisms that are involved in the domain of seismic engineering are discussed. The system has been implemented in PROLOG while procedural tasks are implemented in C. The system is capable of understanding restricted natural language input.

Work towards the development of a knowledge-based system which might be an aid in the management of the safety of a project is described in Chapter 2 by D. I. Blockley. Case histories of projects are collected in the form of event sequence diagrams into a hierarchically structured set of concepts and relations. An open-world mathematics of interval probability as a theory of evidential support is introduced.

T. Krauthammer in Chapter 3 puts emphasis on the need to combine symbolic and numeric approaches in order to assess realistically structural damage following a major disaster, such as a flood or an explosive event. A risk analysis of a concrete gravity dam illustrates the general discussion of engineering evaluation for hazard prevention.

A consultation system for constitutive modelling of materials in solid mechanics is discussed in Chapter 4 by J. Ambroziak and M. Kleiber. Selection of a material model (i.e. of equations describing the material's thermo-mechanical properties) is an important task in correctly posing and solving boundary value problems typical of the non-linear mechanics of deformable bodies. The authors critically discuss features routinely present in existing expert system shells and explain reasons for employing a version of the blackboard framework in this work.

P. Wriggers and N. Tarnow indicate in Chapter 5 that controlling finite element algorithms by expert systems may turn out to be highly beneficial in cases where calculations do not depend entirely on algorithms but also on heuristic data. In the implementation discussed in this work an interface relates the numerical results of finite element computations to the rules of an expert system. Complex non-linear calculations illustrate the effectiveness of the coupled system.

In the closing chapter of Part I of this volume D. Hartmann identifies areas within the computational mechanics field in which expert system applications seem to be promising. A pilot program in the form of a 'solution assistant' for the numerical analysis of plane stress problems is described.

Part II of the book starts with Chapter 7 by K. Schittkowski and R. Zotemantel in which they introduce an interactive programming system that supports the whole life-cycle of the solution process of a mechanical structural optimization problem. Given the geometry of the structure by means of a finite element description, the system guides the user to define an appropriate optimization problem. The system is self-learning and uses rules for proposing a suitable optimization algorithm or remedies in an error situation.

The concept of statistical machine learning has been applied by D. A. Hoeltzel and W. H. Chieng in Chapter 8 to a sample database of non-linear programming problems. Conclusions have been drawn about a type of optimization problem and a corresponding sequence of non-linear programming solution algorithms. A program with the capability of learning from statistical pattern recognition is discussed.

M. L. Das and S. K. Ghosh emphasize in Chapter 9 that heuristics have always been used in multistory building design. In this contribution the authors attempt to formalize the use of some of these heuristics by presenting a workable knowledge-based expert system for selection of the cast-in-place slab type of multistory buildings.

In Chapter 10 P. Hajela and J. L. Chen describe the implementation of an expert system that provides interactive assistance in tasks related to the optimum synthesis of structural systems. These tasks include finite element modelling of the problem domain, building an optimum design model and selection of optimization strategies and parameters for the problem.

P. Y. Papalambros takes as a starting point in Chapter 11 the fact that the overall computational cost in optimal design is dominated by function and gradient evaluations of the constraint functions. This necessitates the use of active set strategies where only a subset of the original constraint set is used for computation in any particular iteration. A framework for developing knowledge-based active set strategies is described, motivated in part by application to aircraft structures.

In the closing chapter, D. Aunay, S. Aunay, D. Chorlay, G. Touzot and M. Vayssade emphasize that the coupling of existing AI and numerical programs implies several difficulties and conflicts. A new software architecture is proposed, which is designed to simplify the implementation of coupled methods. The architecture is based on an object-oriented data manager and on a stretchable set of independent commands. Two specific examples are considered to illustrate the chapter: the first deals with an intelligent user interface and the second with shape optimization under technological constraints.

The book should attract the interest of researchers, graduate students and innovative practitioners who employ computer techniques in the fields of mechanical, aeronautical and civil engineering and design. Computer scientists will find in the book a useful overview of several up-to-date applications of AI.

Michał Kleiber

PART I
Assessment of Structural Behaviour

1

QSEIS: a system for deriving qualitative seismic behaviour of structural systems from structural descriptions

J. Ganguly, E. Kausel and **D. Sriram**
Massachusetts Institute of Technology

1. INTRODUCTION

The analysis and design of complex or critical facilities, such as tall buildings, nuclear power plants, offshore structures, or cable-stayed bridges, requires that these facilities be made safe against the effect of earthquake-induced loads in seismic areas. The various stages involved in the design of these facilities are as follows:

(1) *Problem identification.* The problem, necessary resources, target technology, etc., are identified.
(2) *Specification generation.* Design requirements and performance specifications are listed.
(3) *Concept generation.* The selection or synthesis of potential design solutions, such as a structural system, is performed. Several alternative designs may be proposed.
(4) *Analysis.* The response of the system to external effects is determined by means of appropriate models.
(5) *Evaluation.* Solutions generated during the concept generation stage are evaluated for consistency with respect to the specifications. If several designs are feasible, then an appropriate evaluation function is used to determine the best.
(6) *Detailed design.* The components of the chosen system are such that all applicable constraints (or specifications) are satisfied.
(7) *Design review.* The detailed design is checked for global consistency.

There may be significant deviations between the properties of components assumed or generated at the concept generation stage and those determined at the detailed design stage, which would necessitate a reanalysis. The process continues until a satisfactory or optimal design is obtained.

Currently, several sophisticated computer programs exist that allow us to perform analysis and design of complex structures. However, none of these programs has any reasoning capability. Users of numerical packages must anticipate the behaviour prior to analysis. The purpose of the analysis is usually to confirm the anticipated behaviour. Furthermore, the responsibility of validating the computer outputs lies entirely with the user.

A number of human mental activites, such as developing computer programs, performing symbolic mathematical manipulations, engaging in commonsense reasoning, processing natural language, making engineering decisions, are all said to require 'intelligence'. Computer scientists have been fairly successful in developing prototypical computer programs for performing the above-mentioned tasks. One may say that such systems possess artifical intelligence (AI) to some extent. Ultimately the goal of AI is to develop an information processing theory of intelligence that can be used to design intelligent machines as well as explicate intelligent behaviour of human beings.

In the recent past, AI techniques in the form of knowledge-based expert systems (KBES) have begun to emerge from computer science laboratories into industrial settings. KBES are computer programs that use knowledge in a given field (formalized in a knowledge base) as well as inference procedures to address problems whose solutions require significant human expertise (an overview of KBES is provided in [20]). The KBES technology has been applied successfully to a wide variety of problems ranging from diagnosing diseases, evaluating potential mineral deposits, identifying structures of complex organic materials to designing buildings, VLSI chips, etc. (an extensive bibliography in engineering is provided in [20]). A major difficulty in developing KBES is in representing and using knowledge that human experts use in problem solving. This problem is compounded by the fact that human expertise is often empirical and even imprecise. Most KBES such as SACON [2,3], which was developed as an intelligent front end to a finite element analysis program, rely on a technique known as rule-based deduction. In such systems, the expert knowledge is represented as a large set of rules, and these rules are used to guide a conversation between the system and the user and finally to deduce a conclusion. Hence, current KBES are limited because they have little or no knowledge of the underlying physics of a problem. In addition, these KBES cannot reason about the behaviour of physical systems in a qualitative manner, as experienced engineers normally do.

In this paper, we describe QSEIS, a KBES, that can provide expert advice on the seismic behaviour of structural systems, based on heuristic expert knowledge and qualitative reasoning capabilities. We discuss the nature of knowledge structures and reasoning mechanisms that are involved in the domain of seismic engineering (section 2). QSEIS consists of a layered knowledge base and a friendly interface (section 3). The knowledge base consists of heuristic rules as well as more fundamental knowledge or laws based on principles of mechanics. The heuristic knowledge has been implemented as rules while an attempt has been made to implement the more fundamental laws of mechanics as causal relationships between entities (section 4). The user interface consists of a diagram interpretation module and a natural language interpreter (section 5). QSEIS accepts the following types of inputs: menu driven, restricted natural language and structural sketches. It writes the derived behaviour in a format that can be processed by the Latex document processor. QSEIS has been implemented in PROLOG; procedural tasks are implemented in C (section 6). Examples of input and output are presented (section 7).

2. ENGINEERING PROBLEM SOLVING

To appreciate the various reasoning mechanisms that are involved in an actual engineering problem (specifically in the domain of seismic engineering), let us consider the structural system as shown in Fig. 1.1. From the drawings in Fig. 1.1 an experienced engineer might reason as follows, even before he performs any detailed calculations:

> The height-to-depth ratio of the structure seems to be quite large and therefore there is a possibility of large overturning moments. The shear walls have large openings and hence will behave as frames. This implies that the structure will primarily behave as a frame and therefore is quite flexible. A heuristic rule for obtaining the fundamental period (T) of a structure that acts as a moment frame is $0.025 \times h^{3/4}$, which for the present structure is about 1.8 seconds ($0.025 \times 300^{3/4}$). Since a frame behaves as a shear beam, the higher modes are likely to correspond to time periods which are approximately $T/3=0.6$ seconds, $T/5=0.36$ seconds, etc. Furthermore, since the plan is almost regular, it is sufficient to perform modal analysis and consider modes of vibration corresponding to 1.8 and 0.6 seconds to estimate forces and deflections. The overturning moment is likely to be higher than usual since the lateral distribution will be non-linear instead of being almost linear. The slight non-symmetry in the plan will cause some lateral forces which in turn will cause torsion. Since the shear walls have large openings, there is a possibility of serious degradation of performance due to the torsion. The column in the first story are very tall and this would cause very large drift in the first floor which might cause non-structural damage. The coupled shear walls are not identical in nature, i.e. they possess different stiffnesses. This can cause large forces in the links and since the links are not restrained laterally, there is a serious possibility of their buckling. ...

The above example suggests the following types of knowledge structures and reasoning mechanisms that the engineer utilizes:

(1) *Heuristic knowledge,* such as

      if the building behaves as a moment frame, then the fundamen-

18 ASSESSMENT OF STRUCTURAL BEHAVIOUR [Pt. I

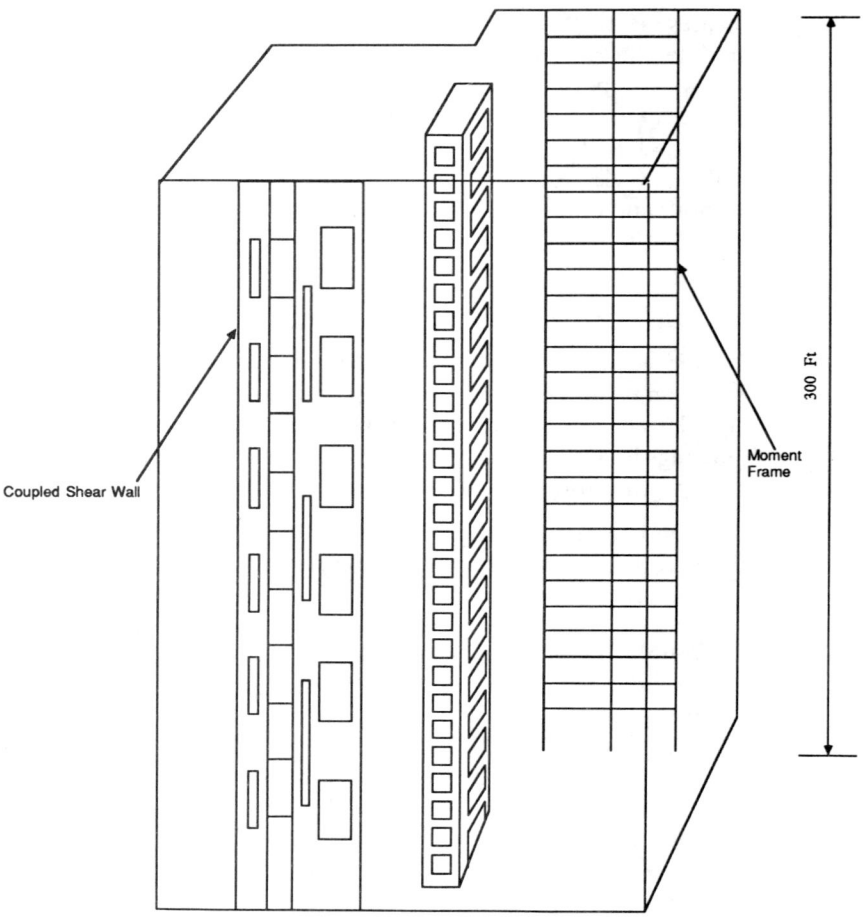

Fig. 1.1 — Example structural system.

tal period of the structure is $0.025 \times 300^{3/4}$ (the height of the building is 300 feet)

(2) *Qualitative analysis,* such as

the slight non-symmetry in the plan will cause some lateral forces which in turn will cause torsion

(3) *Analogical reasoning,* such as

since a frame behaves as a shear beam, the higher modes are likely to correspond to time periods which are approximately 0.6 seconds, 0.36 seconds, etc.

Here an analogy is made between the frame and the shear beam. Since the higher modes of a shear beam are likely to correspond to time periods which are approximately 0.6 second, 0.36 seconds, etc., the modes of the frame will have similar characteristics.

(4) *Approximate quantitative analysis (back-of-the-envelope calculations),* such as:

   calculating stiffness distribution, estimating lateral load distribution, etc.

Hence there is a need to incorporate the above mechanisms in computer programs that attempt to achieve expert problem solving capability in engineering. The overall goal of our current work is to provide a flexible framework that would incorporate several kinds of knowledge structures and inference mechanisms needed for the reasoning process that experienced engineers would follow in assessing the behaviour of structural systems. In particular, we will focus on the assessment of structural behaviour under seismic loading.

3. ARCHITECTURE OF QSEIS

The various modules of QSEIS are shown in Fig. 1.2. These modules are briefly described below.

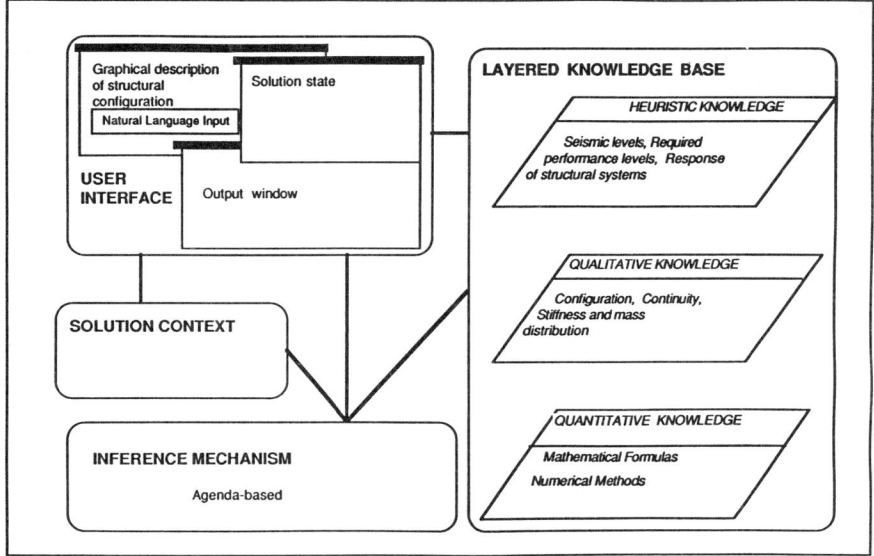

Fig. 1.2 — Modules of QSEIS.

(1) The user interface contains a set of utilities—graphics, menus, restricted natural language, etc.—that facilitate graceful human–computer interaction.

(2) The context consists of the problem description and the solution, which includes the predicted behaviours and the suggested remedial actions.
(3) The knowledge base of QSEIS can be divided into three levels: compiled, qualitative, and quantitative. Production rules, past cases, are encoded at the compiled level. At the qualitative level, the knowledge base consists of causal relationships between the various entities in the domain. The quantitative level normally consists of constitutive, compatibility, equilibrium equations (physical laws), numerical techniques, closed form solutions, etc. There may be several layers at this level. In the current system only approximate analysis techniques, i.e. techniques used by engineers to perform back-of-the-envelope calculations, are incorporated. All three levels are able to access an artifact space, which defines the semantics of various kinds of objects.
(4) The inference mechanism is based on an agenda-based control strategy, where tasks are ordered in an agenda during the problem solving process and the task with the highest priority is executed first.

The overall inferencing strategy in QSEIS can be summarized as follows (see Fig. 1.2).

(1) The user inputs a graphical description of the structure. For example, the plan and the elevation are sketched on the screen. Using a menu interface, the user is able to input other details about the building, such as location, purpose, soil characteristics, etc.
(2) Heuristic knowledge at the compiled level is used to determine whether the structure conforms to some accepted guidelines, such as Applied Technology Council guidelines. If the system finds potential problems, it informs the user. In addition, the system is able to suggest remedial actions.
(3) Although the analysis using compiled knowledge may suggest that the structure does not violate accepted guidelines, there may be some problems that can be detected through commonsense reasoning. This type of reasoning is provided by the knowledge structures at the qualitative level.
(4) Sometimes there will be a need to perform back-of-the-envelope calculations. Appropriate algorithms, residing in the quantitative level, are utilized.

4. KNOWLEDGE BASE

4.1 Compiled knowledge level

The knowledge at this level is comprised of heuristics based on engineering experience and past case histories.

4.1.1 Heuristic knowledge

Reports and guidelines from a number of organizations such as the Applied Technology Council clearly document the effect of the various factors such as shape, size and components on the dynamic behaviour of the structure during an earthquake [1, 5,6,22]. Similarly well-established criteria exist in the literature about the seismic performance of various facilities, risk levels at different sites, ductility characteristics, analysis and design methods, etc. It is necessary to take into consideration all the above factors in order to ensure a proper design. For example, the location of the site decides the effective acceleration, the effective velocity, and the risk level associated with the site. The purpose of the structure and its location together decide the seismic hazard level for which it needs to be designed. In the case of buildings, seismic performance groups have been established depending on the nature of the facility. The seismic performance group to which the facility belongs and the risk level associated with the site together decide the appropriate framing systems, and the analysis and design procedures that are necessary for the facility. The selection of the framing system and the analysis and design procedures are also dependent on the geometric features of the plan, elevation, and the materials used in constructing the facility.

Knowledge (compiled and qualitative) is represented using a collection of frames and rules. Frames are knowledge structures that depict objects, attributes and their values. The relationships between these frames define the semantic network of the domain. A few sample frames are shown below:

structure
    TYPE: building
    PURPOSE: hospital
    LOCATION: California
    VERTICAL ELEVATION: **vertical-elevation-1**
    PLAN: **horizontal-plan-1**

vertical-elevation-1
    HEIGHT: 500 feet
    WIDTH: 100 feet
    SYMMETRY: does not exist
    VERTICAL SETBACKS: exist
    MASS DISTRIBUTION: non-uniform

horizontal-plan-1
    LENGTH: 200 feet
    WIDTH: 100 FEET
    SYMMETRY: does not exist
    RE-ENTRANT CORNERS: present
    DIAPHRAGM STIFFNESS: significant change in level 34
    POTENTIAL FOR LARGE TORSIONAL MOMENT: does not exist

The first frame above indicates that the structure is a hospital building which is located in California. The vertical elevation and horizontal plan of the structure are given by the **vertical-elevation-1** and **horizontal-plan-1** frames respectively.

A rule is a conditional statement linking the values for attributes of objects. Each rule consists of a left-hand side (LHS), i.e. the portion after **if**, and a right-hand side (RHS), i.e. the portion after **then**. The LHS consists of a number of conditions, where each condition should evaluate either to true or false. The RHS consists of actions to be taken if all the conditions in the LHS evaluate to true.

The following are some examples of rules (the acronyms in parentheses are used in the inference network shown in Fig. 1.3); the rules depict only a part of the inference network in Fig. 1.3.

if
   the hazard exposure group of the building is 3 and   (SHEGIII)
   seismic index of location is 4   (INDEX4)
then
   the required seismic performance category is D   (SPCATD)

if
   seismic performance category of the building is D and (SPCATD)
   the height of the structure is greater than 160 feet   (HTGT160)
then
   the appropriate structural system is a dual system or (DUALSYS)
   a special moment frame system   (SPMOMFR)

if
   seismic performance category of the building is D and (SPCATD)
   horizontal plan of structure is irregular   (IRRPLCONFIG)
then
   it is necessary to perform a three dimensional   (DYNCON)
   dynamic analysis

if
   seismic performance category of the building is D and (SPCATD)
   horizontal elevation of structure is irregular and   (IRRELCONFIG)
   the plan of the structure is regular   (REGPLCONFIG)
then
   it is necessary to perform at least a modal analysis   (MODAN)
   of the structural system

Fig. 1.3 depicts part of the rule base in the form of an inference network. The top level hypotheses in the sample inference network determine whether an appropriate structural framing system has been selected. Although Fig. 1.3 illustrates part of a knowledge based related to buildings, the issues that influence the seismic behaviour are general in nature. The knowledge structures used in QSEIS are sufficiently general to address other constructed facilities such as bridges, dams, etc.

4.2 Qualitative level

In addition to well-defined heuristics, there are a number of qualitative issues that play a significant role in the behaviour of a structure during an earthquake. While many of these may seem to be obvious, and are rather clear to an experienced engineer, it is rather difficult to implement such reasoning mechanisms in a computer program. Many of these qualitative

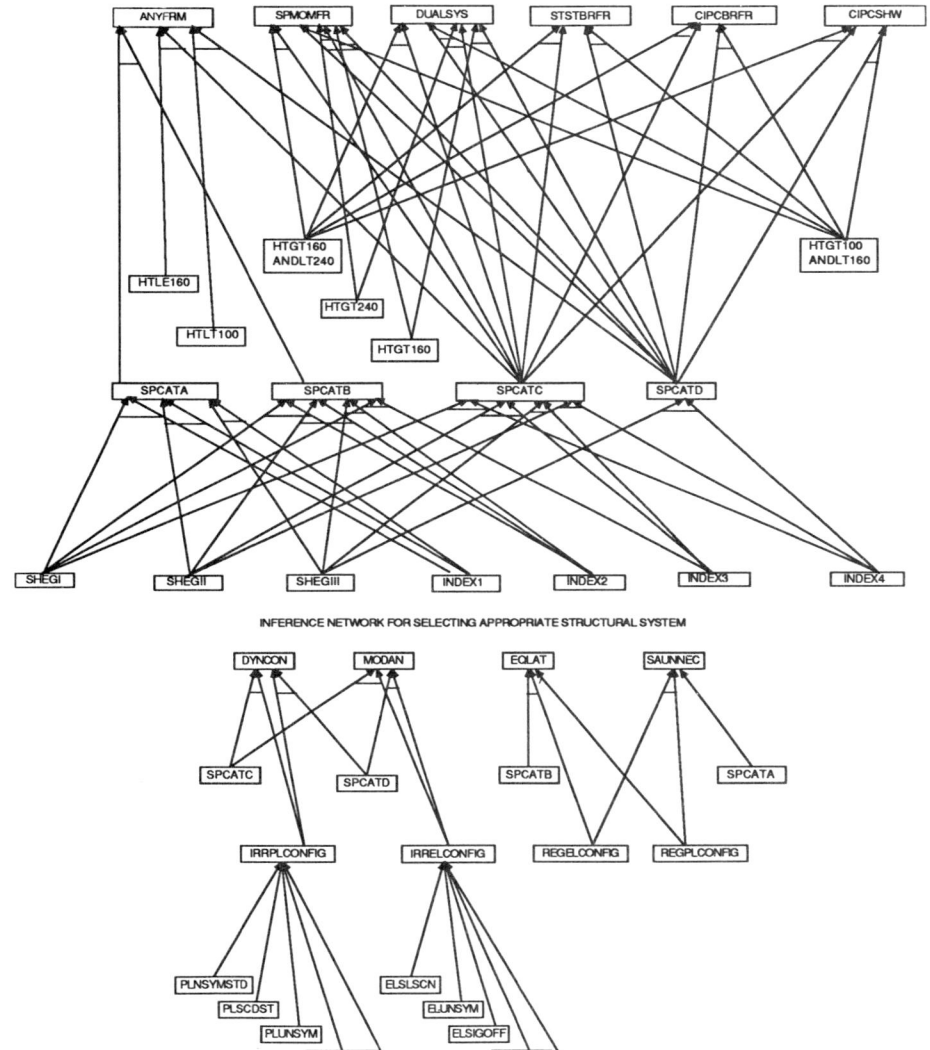

Fig. 1.3 — Inference networks.

issues can be thought to require 'engineering judgement', which is very difficult to model and implement once a domain becomes sufficiently large and complex. The importance of this apect of knowledge becomes apparent when we consider, for example, the effects of the plan shape on the dynamic behaviour of the structural system. If the plan is too long, it is possible that differential movements can cause disastrous effects. Similarly, re-entrant corners may cause significant torsional forces. While these may seem obvious to us, it is not easy to extract such information from structural drawings and infer appropriate conclusions in a general sense.

QSEIS derives the abstract qualitative behaviour using causal and dependency links. The causal knowledge is organized in terms of nodes and links. Nodes can be thought to be clusters of information that describe a particular state. A causal link specifies the cause–effect relation between the cause (antecedent) and the effect (consequent) states. For example, in Fig. 1.4 two planes — the dependency plane and the causal plane — are shown. The dependency plane depicts dependency relationships between various entities. For example, vertical elevation configuration is considered to be irregular if there are significant setbacks (reverse and vertical) and it is considered to be slender if the aspect ratio is greater than five. The causal plane depicts the causal associations between entities (or objects). The vertical lines between the planes indicate that the entity (object) in the dependency plane has some behavioural characteristic associated with it in the causal plane. For example, ECCENTRICITY in the dependency plane is associated with a value LARGE in the causal plane.

For the example described on page 27, a few of the following causal associations (depicted pictorially in Fig. 1.4) are utilized for assessing behaviour:

(1) A rapid change of curvature causes stress concentration.
(2) Openings in structural elements cause reduction in their stiffness.
(3) Deflection due to bending may cause rotational inertia of masses to be considered.
(4) Significant reduction in stiffness causes large deflections.
(5) Non-structural elements cause increase in stiffness of the overall structure.
(6) An increase in stiffness causes reduction in fundamental period of response of the structure.
(7) A reduction in fundamental period will cause increased lateral force.
(8) An increase in fundamental period will cause higher modes of vibration to participate.
(9) Higher modes will cause non-linear distribution of lateral force.
(10) A non-linear distribution of lateral force will cause force distribution to increase with height.
(11) A higher force near the top will cause larger overturning forces.
(12) Large deflections cause non-linear behaviour.
(13) Stiffness distribution causes appropriate force distributions and deflections.
(14) Significant differences in horizontal displacements cause large axial forces in members.
(15) Large axial forces cause buckling of members if they are not laterally restrained.

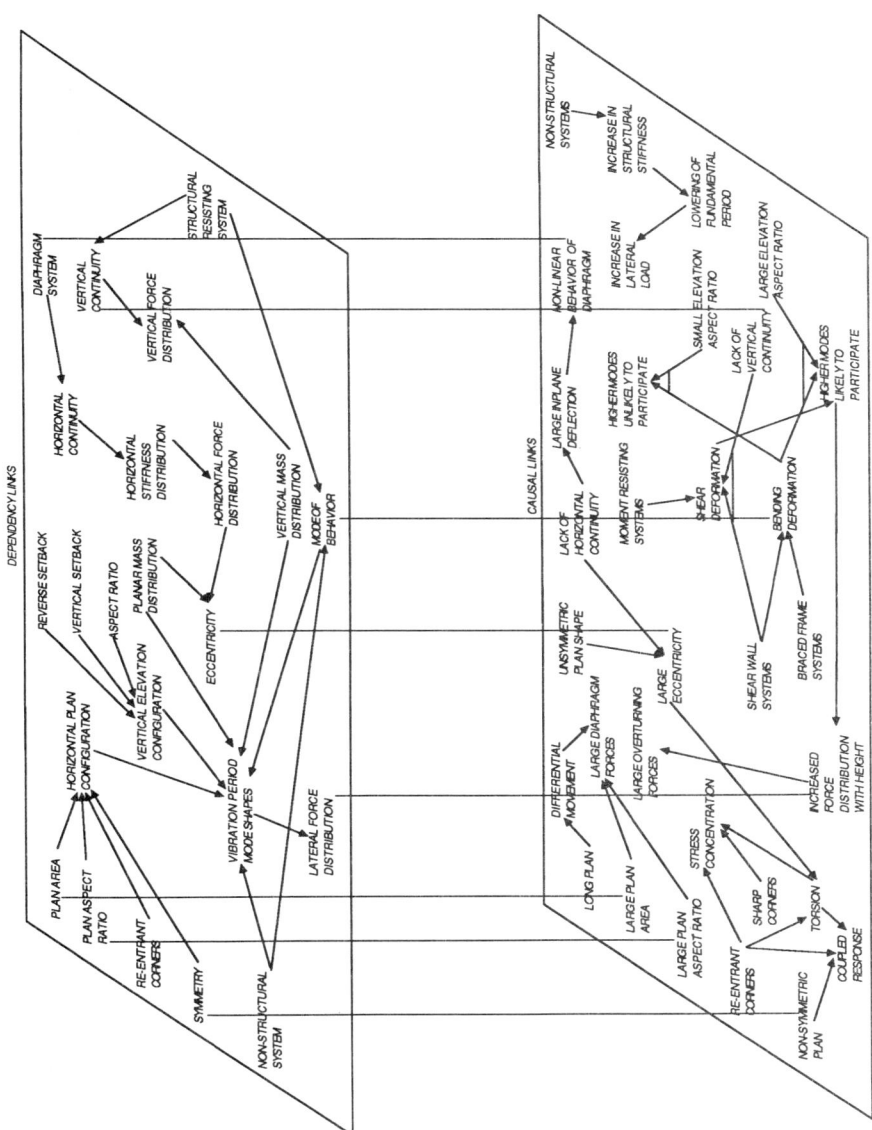

Fig. 1.4 — The dependency plane and the causal plane.

5. USER INTERFACES: DIAGRAM UNDERSTANDING AND GENERATION

User interfaces play a very important role in the acceptance of computer-aided software by engineers. QSEIS provides a sophisticated menu-driven interface that is easy to use. Furthermore, QSEIS can interpret sketches describing plan, elevation and structural systems. The user typically draws these entities using an input device such as a mouse. However, in cases where it is not possible to infer the characteristics of the sketch, it is necessary to inform QSEIS explicitly about the specifics of the sketch. For example, if the user draws a rectangle on the screen, it is not clear whether it is a plan view or an elevation view. Therefore, it is necessary to tell QSEIS about the particular label of this sketch. The interpretation of sketches consists of the following:

(1) Determining features of plans and elevations, such as re-entrant corners, shapes, vertical setbacks and symmetry.
(2) Identification and location of structural systems such as shear walls and moment frames.

A consistent interpretation of the drawing is then used to identify the features that influence the structural behaviour.

QSEIS also allows restricted natural language input. A QSEIS incorporates definite clause grammars (DCG) for language analysis which allows a restricted conversation [17].

6. IMPLEMENTATION DETAILS

QSEIS has been implemented in PROLOG and C and currently runs on the Tektronix 4400 series workstations. All declarative knowledge has been represented as facts and rules in PROLOG; we believe that PROLOG is a good language for rapid development of a prototype that essentially requires a backward chaining mechanism. Procedural tasks such as three-dimensional graphic representation has been implemented in C. PROLOG is similar to a backward chaining, rule-based production system with very specific control information about what to do next. It is a practical and efficient implementation of many aspects of intelligent program execution, such as non-determinism, parallelism, and pattern-directed procedure call. It seems that human beings communicate in declarative languages rather than in programming languages [13], and therefore it is necessary that a programming language allow easy representation of declarative knowledge. PROLOG databases satisfy this criterion to some extent by allowing arbitrary Horn clauses that include free variables. The advantage of declarative information is one of generality. However, universal generalization cannot be done without more predicate calculus than PROLOG allows.

7. EXAMPLE

In this section a number of screen dumps from a sample run are presented to illustrate the ideas that have been discussed earlier. The screen dumps are sequentially arranged and illustrate the following:

(1) Fig. 1.5 shows the starting banner of QSEIS.
(2) Fig. 1.6 illustrates the menu-driven input with the explanation facility available in QSEIS. In this case, the system asked the user to identify the soil profile at the site. The user responded by asking the system to explain the question further, such as why the question was being asked and what the different alternatives in the menu meant.
(3) Fig. 1.7 illustrates a plan shape that the user has sketched using a mouse.
(4) Fig. 1.8 illustrates the three-dimensional view of the final structural system generated from user-provided sketch of the plan, elevation and structural system.
(5) Fig. 1.9 displays the inference process at execution time. Here the system deduced that the plan is not regular since the plan shape is not symmetric although the corners of the plan are not re-entrant and the floor mass distribution is uniform. The white boxes imply that the appropriate clauses have been found to be true, whereas the black boxes imply that its contents have not been found to be true. The tilde sign is a symbol for 'not'.
(6) Fig. 1.10 displays a particular state during program execution and causal reasoning.
(7) Fig. 1.11 illustrates an example of a question that may be posed to QSEIS by the user in a restricted natural language. The adjacent window illustrates part of the parsing–interpretation process.
(8) Fig. 1.12 illustrates the answer presented by QSEIS.
(9) Finally the output from QSEIS, describing the derived behaviour from user provided information and sketches, is provided.

QSEIS report: qualitative behaviour under seismic loading
    The site is located in kern county in the state of California. According to the seismic zoning maps of USA, the map area identification numbers for peak acceleration and velocity are 7 and 7 respectively. Hence the coefficients corresponding to the peak seismic accelerations and velocities of this site location are estimated to be 4.0E-1 and 4.0E-1 respectively. The seismicity index of the location is rated to be 4 and because the hazard exposure group of the structure is 2, the structure is classified as belonging to seismic performance category c.
The seismic performance category of this structure is c, the height is greater than 240 feet, hence the preferable seismic structural resisting systems are moment resisting frame system with special moment frames or dual systems or structural steel or cast in place concrete braced frames or shear walls.
The number of openings in the diaphragm is not__many, the size of the openings in the diaphragm is not__large, hence the diaphragm can be characterized to be strong. The length of the plan is not__long, hence relative lateral displacement of the diaphragm is not__likely.
The length of the plan is not__long, the strength of the diaphragm is strong, the

QSEIS: A Framework for Deriving Qualitative Seismic Behavior from structural descriptions

developed by
Jaideep Ganguly

under the supervision of Professors
Eduardo Kausel and Duvvuru Sriram

Intelligent Engineering Systems Laboratory
Department of Civil Engineering
Massachusetts Institute of Technology

Fig. 1.5 — QSEIS screen dumps: the starting banner.

differential movement of the diaphragm is not_likely, hence the stresses in the diaphragm are likely to be not_large.
The corners of the plan are reentrant, the plan shape is not symmetric about one axis, the diaphragm is considered to be strong, the floor mass distribution is uniform, hence the plan configuration of the structure can be considered to be irregular.
Setbacks in elevation are not_present, reverse setbacks in elevation are not_present, the vertical mass distribution is not_uniform, hence the elevation configuration of the structure is considered to be irregular.
The seismic performance category of the structure is c, the plan and elevation configurations are irregular and irregular respectively, hence the suggested analysis procedure is three_dimensional_analysis. The design story drift based on an elastic analysis should be amplified by 5.
The vertical resisting system of the structure includes Reinf. Conc. Shear Walls and Braced Frames, and the number of openings in the shear wall is none, hence the structure will deflect primarily due to bending.
The material of the vertical resisting system is concrete.
The structure will deflect primarily due to bending, the fundamental period of response is likely to be about 2 seconds. The fundamental period is between 0.5–2.5 seconds, hence the seismic load distribution will be between linear and parabolic.
The corners of the plan are reentrant. The shape of the plan is not symmetric about one axis, hence the plan is likely to be subjected to torsion.
The plan of the structure will be subjected to torsion, hence the plan will experience stress_concentration.
The diaphragm will be subjected to torsion, hence the dynamic response of the structure will be coupled. The aspect ratio of the elevation is not_large, hence the number of higher modes likely to participate in the dynamic response is not_large.
The seismic lateral load is estimated to be 7 percent of the weight of the structure.
Non structural systems are present, the stiffness of the structure will be increased, fundamental period of the structure will be decreased, hence the lateral load on the structure will be increased.
Tall columns in vertical resisting system are present, soft story effect is likely, stiffness of vertical resisting system is reduced, lateral deflection of vertical resisting system is large, hence P-Delta effect is likely to be significant.

> *Explanation:*
>
> *The effect of site conditions on building response depend on the soil profile types. The seismic wave velocity at the site depends on the soil profile, which in turn affect influence the fundamental period of the soil stratum. If the fundamental period of the stratum and the structure are sufficiently close, soil structure interaction is likely to be significant.*
>
> *On the basis of the available body of data, the four conditions were selected as follows:*
>
> *1. Rock — of any characteristic whether it is shalelike or crystalline in nature. As a general rule, such material is characterized by a shear wave velocity of greater than about 2500 fps.*
>
> *2. Stiff soil conditions or firm ground — including any site where soil depth is less than 200 feet and the soil types overlying rock are stable deposits of sands, gravels, or stiff clays.*
>
> *3. Deep cohesionless or stiff clay soil conditions — including sites where the soil depth exceeds about 2500 feet and the soil types overlying rock are stable deposits of sands, gravels or stiff clays.*
>
> *4. Soft-to-medium stiff clays or sands — characterized by sereval tens of feet of soft-to-medium stiff clay with or without intervening layers of sand or other cohesionless soils.*
>
> *In locations where soil properties are not known in sufficient detail to determine the soil profile types or where the soil profile does not fit any of the types listed in the menu, soil profile S2 may be used.*

Please identify the soil profile at the location.	Rock of any type
	Stiff soil conditions
	Deep cohesionless soil
	Soft to medium stiff clay and sand
	Explain
	Quit

Fig. 1.6 — QSEIS screen dumps: menu-driven input and explanation facility.

The soil profile at the site consists of stiff soil conditions, hence the shear wave velocity at the site will be large.
The shear wave velocity at the site will be large, the depth of the soil at the soil at the site is not__large, hence the fundamental period of the soil is not__large.
The period of the structure is not__large, the fundamental period of the soil is not__large, hence soil structure interaction is likely to be significant.
The site is situated on__ridge, hence the wave energy will be concentrated.
The fundamental period of the soil at the site is not__large but the shear wave velocity is large, hence the wave length of the shear wave may be large.
The horizontal extent of the soil is large, the horizontal profile of the soil is

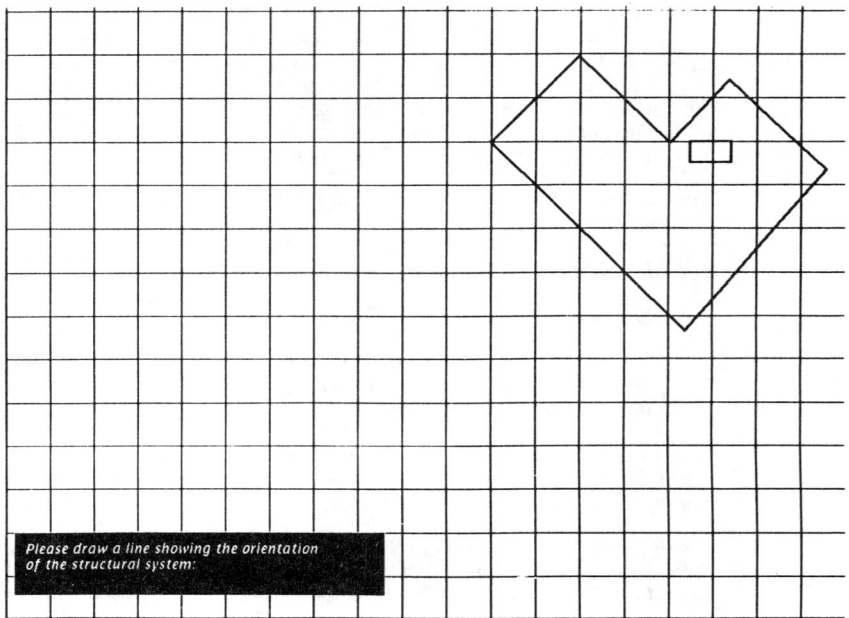

Fig. 1.7 — QSEIS screen dumps: plan shape sketched by user.

not__uniform, hence relative displacement of the structure will occur, hence the structure is not__safe.

The slope of the bedding plane at the site is steep, hence multiple reflections of seismic waves will occur. The depth of the stratum is not__large, the wave length of the shear waves could be large, hence seismic energy buildup is likely, and hence reverberations at the site are likely.

8. FUTURE WORK

Currently we are investigating strategies of knowledge acquisition and inductive learning. Appropriate strategies will be incorporated in QSEIS in future.

ACKNOWLEDGEMENTS

We would like to thank Dr Mike Freiling of Tektronix Inc., for providing the workstations on which QSEIS was developed. We would also like to extend our thanks to Logicware Inc., for providing us with MPROLOG.

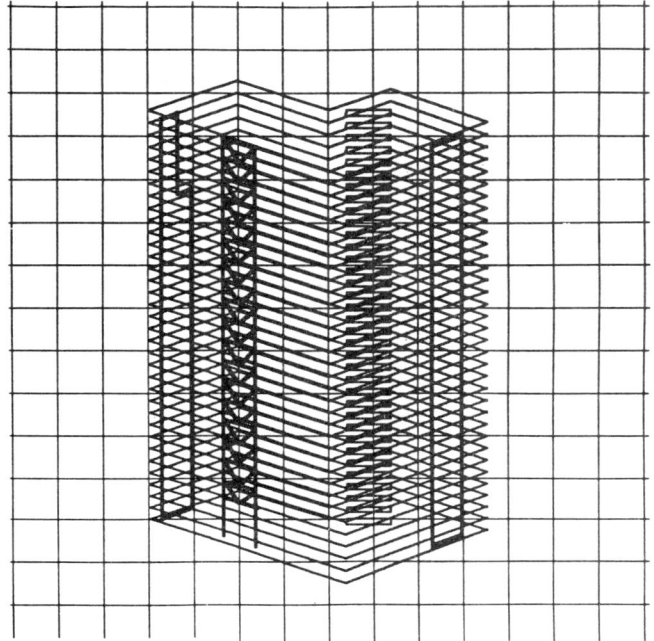

Fig. 1.8 — QSEIS screen dumps: three-dimensional view of final structural system.

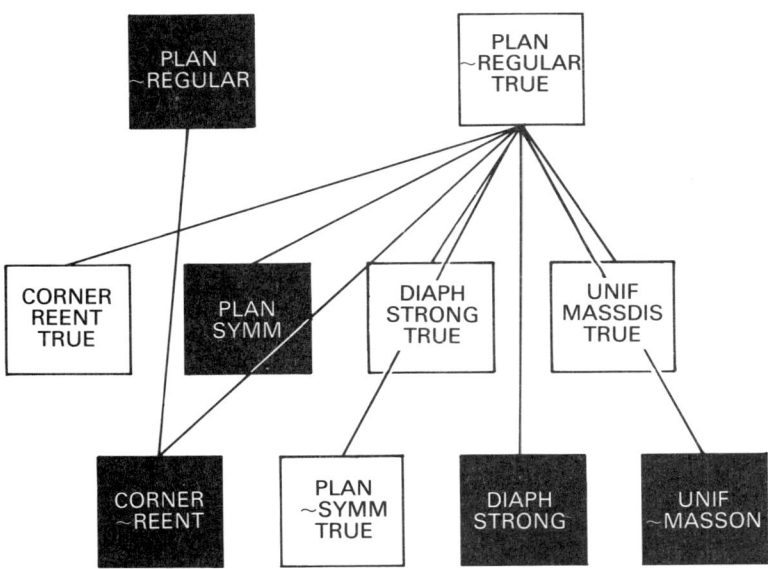

Fig. 1.9 — QSEIS screen dumps: inference process at execution time.

Diaphragm strength:
strong

Floor mass distribution:
uniform

Plan configuration:
irregular

Casual Reasoning:

The size of diaphragm openings is:
notlarge
Hence, the diaphragm is characterized as:
strong

The length of the plan is:
notlong
Hence relative lateral displacement of the diaphragm is:
notlikely

Relative diaphragm lateral displacement is:
notlikely
The strength of the diaphragm is:
strong
Hence, the stresses in the diaphragm is likely to be:
notlarge

The corners of the plan are:
reentrant
The shape of the plan is:
xunsymmetric
ysymmetric
The strength of the diaphragm is
strong
The floor mass distribution is:
uniform
Hence the plan is considered to be:

Please click a button to continue . . .

Fig. 1.10 — QSEIS screen dumps: particular state during program execution and causal reasoning.

> *Restricted Natural Language Interpretation:*
>
> *Key Word Group: <KW>1*
> *Phrase Structure Pattern: <NP>8*
> *Phrase Structure Pattern: <VP>10*
> *Phrase Structure Pattern: <VP>31*
>
> *Trying to Satisfy Grammar Rule: 4:1*
> *Phrase Structure Pattern: <VP>8*
> *Key Word Group: <KW>1*
> *Phrase Structure Pattern: <NP>2*
> *Key Word Group: <KW>2*
> *Phrase Structure Pattern: <NP>2*
> *Key Word Group: <KW>1*
> *Key Word Group: <KW>2*
> *Key Word Group: <KW>1*
> *Key Word Group: <KW>2*
> *Key Word Group: <KW>1*
> *Phrase Structure Pattern: <NP>8*
>
> *Trying to Satisfy Grammar Rule: 4:2*
> *Phrase Structure Pattern: <VP>8*
> *Key Word Group: <KW>1*
> *Phrase Structure Pattern: <NP>2*
> *Key Word Group: <KW>2*
> *Phrase Structure Pattern: <NP>2*
> *Key Word Group: <KW>1*
> *Key Word Group: <KW>2*
>
> *Please click a button to continue . . .*

> *Please type in your statement now.*
> *If you do not have any statement, type:*
> *<exit>*

*How does the fundamental period of the building influence the design of the structure?

Fig. 1.11 — QSEIS screen dumps: question asked by user in a restricted natural language.

> *Answer:*
>
> *Structures whose fundamental periods are long should be designed more conservatively as compared to structures whose periods are short. The fundamental period of the building increases with the number of stories. Hence, the longer the period, the larger the likely number of stories and therefore the number of degrees of freedom: hence, the more likely that high ductility requirements can be concentrated in a few stories of the structure, at least for some earthquakes. Also the number of potential modes of failure increases, generally with the period. If the design spectra were proportional to the response spectra for single degree of freedom systems, the probability of failure would increase with the period. Furthermore the instability of the structure is more of a problem with increasing period.*

> *Do you agree with the answer?*
> *If not, you may change the answer. The modified answer will be stored for future reference.*

Yes
No
Explain

Fig. 1.12 — QSEIS screen dumps: answer presented by QSEIS.

REFERENCES

[1] Applied Technology Council, *Tentative Provisions for the Development of Seismic Regulations for Buildings,* 1978.
[2] Bennet, J. S., Creary, L., Engelmore, R. and Melosh, SACON: a knowledge based consultant for structural analysis, *Report STAN-CS-78-699,* Stanford University, September 1978.
[3] Bennet, J. S., SACON: a knowledge based consultant for structural analysis, *Proceedings of 6th IJCAI, August 20–23, 1979,* pp. 47–49.
[4] Blevins, R. D., *Formulas for Natural Frequency and Mode Shape,* Van Nostrand Reinhold, New York, 1979.
[5] Dowrick, D. J., *Earthquake Resistant Design,* Wiley, New York.
[6] Fintel, M. (ed.), *Handbook of Concrete Engineering,* Van Nostrand Reinhold, New York, 1974.
[7] Forbus, K., Qualitative process theory *Artif. Intell.,* **29**, 85–168, 1984.
[8] Genesereth, M. R. and Nilsson, N. J., *Logical Foundations of Artificial Intelligence,* Morgan Kaufmann, Los Altos, CA.
[9] Hayes, P. J., In defense of logic, *Proceedings of 5th IJCAI, 1977,* pp. 565–569.

[10] Kowalik, J. S., *Coupling Symbolic and Numerical Computing in Expert Systems,* North-Holland, Amsterdam, 1986.
[11] Kuipers, B., Qualitative simulation, *Artif. Intell.,* **24**, 1984.
[12] Lenat, D., Prakash and Shepherd, M., CYC: using common sense knowledge to overcome brittleness and knowledge acquisition bottlenecks, *AI Mag.,* 65–85, Winter 1985.
[13] McCarthy, J., Generality in artificial intelligence, *Commun. ACM,* **30** (12), 1030–1035, 1987.
[14] Minsky, M. A., *A Framework for Representing Knowledge, Psychology of Computer Vision,* McGraw-Hill, New York, 1975.
[15] Nilsson, N. J., *Principles of Artificial Intelligence,* Tioga Publishing Company, Palo Alto, CA, 1980.
[16] Patil, R., Causal representation of patient illness for electrolyte and acid–base diagnosis, *Report MIT/LCS/TR-267,* Laboratory for Computer Science, Massachusetts Institute of Technology, October 1981.
[17] Pereira, C. N. F. and Warren, D. H. D., Definite clause grammars for language analysis — a survey of the formalism and a comparison with augmented transition networks. In Grosz, B. J., Jones, K. S. and Webber, B. L. (eds), *Readings in Natural Language Processing,* Morgan Kaufmann, Los Altos, CA.
[18] Raiman, O., Order of magnitude reasoning, *Proceedings 5th NCAI (AAAI-86),* August 11–15, 1986, Morgan Kaufmann, Los Altos, CA. pp. 100–104.
[19] Simmons, R., The use of qualitative and quantitative simulations, *Proceedings 3rd NCAI (AAAI-83),* Morgan Kaufmann, Los Altos, CA. 1983, pp. 364–368.
[20] Sriram, D. (ed.), *Computer Aided Engineering: the Knowledge Frontier,* IESL, Department of Civil Engineering, Massachusetts Institute of Technology, 1988.
[21] Sriram, D., Maher, M. L. and Fenves, S. J., Knowledge-based expert systems for structural design, *Comput. Struct.,* 1–9, January 1985.
[22] Wiegel, W. (ed.), *Earthquake Engineering,* Prentice-Hall, Englewood Cliffs, NJ.
[23] Winston, P., *Artificial Intelligence,* Addison-Wesley, Reading, MA, 1984.
[24] Yip, K. M., Extracting qualitative dynamics from numerical experiments, *Proceedings 6th NCAI (AAAI-87),* Morgan Kaufmann, Los Altos, CA, 1987, pp. 665–670.

2

Knowledge-based systems in engineering risk control

D. I. Blockley
University of Bristol

1. INTRODUCTION

Are engineering failures inevitable? Are they the result of technical errors of some kind of human mistakes or are they just 'Acts of God'? Clearly individual cases merit individual enquiry to establish what went wrong and the lessons to be learned. However, a general conclusion which has emerged from research studies on a number of failures is that human error is a major factor.

One of the most important tasks of the engineering profession is to study the past systematically and to attempt to learn from these failures. Indeed, an engineered structure can be viewed as an expression of current design and construction practice — as a conjecture — and its failure as a refutation of part of that conjecture.

Civil engineers are increasingly aware that the methods of artificial intelligence may have something to offer them. Phrases such as knowledge engineering, knowledge bases and expert systems are being used, and the merits of computer languages such as LISP and PROLOG are being discussed. A large number of popular texts (e.g. [1,2]) explore these basic ideas and present simple applications of AI. A number of researchers are exploring the possibilities in civil engineering [3–5] largely following on from applications in other fields [6,7].

The term expert system is a misnomer, however, since no computer program can be held legally responsible for its recommendations. An engineer is an expert to the extent that he is a responsible decision maker. These computer programs should therefore be called 'advice systems' to avoid misunderstanding between the designers and the users of such systems. An engineer who uses the system should think of it as a means of obtaining further advice about a problem. The advice will not be perfect and may in fact be considerably lacking in completeness. The advice should be considered by the engineer together with other sources of information and advice to help him to come to an engineering decision. Therefore, all advice systems should be able to be interrogated and should be able to expose the reasoning for a particular conclusion.

An advice system contains a number of modules, the two most important being the knowledge base and the inference mechanism. The other modules deal with such matters as explaining how decisions are reached and providing a friendly user–machine interface. The knowledge base contains facts and rules which represent the expert view. A rule states that something is true or dependable or will happen if certain conditions are satisfied. Of course, the rule may not always apply and its concepts may not be precisely defined. The uncertainty of application may, for example, be modelled using probability theory, but there may be doubts about the actual values of the probabilities concerned, particularly if relevant facts cannot be assumed to be independent. The truth of the conditions in the rule may be uncertain. Other rules and facts may be used to provide evidence to support a conclusion. Some of this evidence may be contradictory. For example, some may support the conclusion and some may stack up against the conclusion. Also, there is no real reason to expect that these probabilities must total 1. A lack of evidence supporting a conclusion does not necessarily result in its negation. On the contrary, the lack of evidence may arise because information is hidden, obscure or ill defined. For example, to decide whether something is a bush or a tree may be difficult not because only part of the object can be seen but because it is difficult to be precise about the definition of a bush or a tree.

2. RELIABILITY OR RESPONSIBILITY

The deterministic treatment of engineering calculations has its roots in the ideal of 'exact' science. This conception of science as leading to the 'truth' has been destroyed. The prevailing scepticism amongst philosophers about the possible achievements of science has to be contrasted with the apparent success of technology. A number of recent philosophers have tried to resolve this problem of science. For example, Lakatos argues that the scientist aims at a cumulatively fruitful research programme. Kuhn is perhaps more concerned with the sociology of the changes within science which lead to new theories' being accepted. Feyerabend argues the rival theories incommensurable and no common sets of values for arbitration between them exist. Carnap tried to replace the lost certainty of the last century by a measure of justification based on mathematical probability. He argued that it is meaningful to talk of the probability of the truth of a proposition over a finite and determinate interval. The use of this idea together with the interpretation of mathematical probability as a degree of belief (as argued by de Finetti for example) has been the basis of decision theory and the initial treatment of system uncertainty in reliability theory.

The clash between those following Carnap and those following Popper has been amply discussed by Lakatos. He showed that there is a basic confusion between the notion of the use of probability theory as a 'rational betting quotient' and as a 'degree of evidential support'. Cohen has also destroyed the idea of using mathematical probability as a measure of dependability of inductive evidence and has suggested a new inductive probability measure, quite different from mathematical probability, for application to legal questions.

It is suggested that what really matters to an engineer is the dependability of a proposition. Of course if a proposition is true it is dependable but if a proposition is dependable it is not necessarily true. Truth is a sufficient condition but not a necessary condition for dependability. Einstein demonstrated that Newtonian mechanics is not 'true' but is dependable under certain conditions. Repeatable testing of propositions deduced from Newton's laws has shown that they correspond (within defined error bounds) to the facts (are true) but not always. In other words, the truth content of Newtonian mechanics is high even though in a strict sense the laws are false. They are highly tested, highly corroborated and therefore inductively very reliable but the logical probability of their truth is zero. In solving an engineering problem a whole hierarchy of theories may be used and each theory is only applicable under certain conditions (e.g. elastic behaviour of a material). Even under those conditions the dependability of the use of the theory may not be high (e.g. elastic behaviour of a subsoil) and it is this that constitutes part of the system uncertainty. The sufficient conditions for dependable information have been discussed in detail [8]. A conjecture is dependable if (i) a highly repeatable experiment can be set up to test it, (ii) the resulting state is clearly definable and repeatable, (iii) the value of the resulting state is measurable and repeatable, and (iv) the test is successful. These are sufficient, but not necessary, conditions because the proposition may not be false even though it is not possible to set up repeatable experiments. Deficiencies in any of the ways in which the propositions can be tested or inductively applied obviously lead to uncertainty and a consequent loss of dependability.

If the engineer cannot rely even on science to provide the 'truth' the problem of ensuring an adequate product seems to become overwhelmingly difficult. Engineering judgement is required not only for the matching between engineering theories and the actual product, but also for the assessment of the dependability of the theory iteslf. Settle [9] has tried to resolve this problem by suggesting another criterion, which has great appeal because it is in effect adopted by engineering practice. His sugggestion is made in the light of the critical method in science, which is the development of Popper's philosophy of trial and error, of conjecture and refutation, the concept of critical discussion leading to progress. The notion of the inductive reliability of a theory or hypothesis, argues Settle, should be replaced by the notion of a responsibility to act on the theory of hypothesis. The taking of responsibility implies not that one has earned the right to be right, or even nearly right, but that one has taken what precautions one can reasonably be expected to take against being wrong. The responsible engineer or scientist is not expected to be right every time, but he is definitely expected never to make childish or lay mistakes.

Engineering involves decision making on the basis of information of varying inductive applicability or dependability. A decision may be viewed as a suspension of criticism for a moment. Good decision making is not a static process, however, and criticism of the consequences must therefore be continued after the moment of the decision. Responsibility, it is argued, is a more useful concept than reliability. In one sense this proposition is acceptable because it points directly to the role of the individual and his duty to work with care and diligence. It is a concept used in law where under the law of tort, for example, the standard of care is that of the reasonable practitioner or one who is careful, informed and self-critical. In another sense, however, the concept of responsibility is difficult to accept. It is difficult to define precisely and it varies very much with individual circumstances. A person holding himself out as possessing a particular skill (a university researcher undertaking specialist consulting work) must exhibit the degree of knowledge and skill reasonably to be expected of any other similar person, but he must not be expected always to exhibit the very highest degree of skill, or to anticipate and avoid every possible future risk inherent in the particular task he is carrying out.

The definition of what is reasonable action must in the end be made by the peer group of the person concerned. For the average engineer, one peer group is the membership of the appropriate professional institution. For the detail of a design procedure, a particular code of practice may be the standard of reasonable design practice and must therefore be interpreted as a measure of peer group opinion.

3. OBJECTIVES

The aim of the work described in this chapter is 'to produce a knowledge based computer system which might be an aid in the management of the safety of a project' [12]. The central points raised in attaining this objective are (a) the examination of case histories, (b) the need to handle open world uncertainty, (c) the use of hierarchically structured knowledge base for 'appropriate' problem solving, (d) the interpretation of measurements taken from civil engineering systems.

The first problem to be addressed is that of obtaining accurate details of projects, both failed and successful, past and present. Information is combined from two sources — published reports and recorded interviews. Both have their disadvantages. Written reports commonly follow inquiries into major failures, e.g. Tacoma Narrows Bridge or Ronan Point. While generally providing a wealth of detail, they represent only one extreme of the continuum of failures. Much more significant in terms of numbers of occurrences are the small-to-medium failures which may not undergo detailed and public investigation. The potential information which could be learned from these is lost to the industry. A second approach which has been used with some success is interviews with key individuals concerned with a project [10,11]. This enables information to be gleaned from cases which may not otherwise have been reported. However, as Blockley [12] has noted thare is a narrow 'window' of time during which suitable case studies are available. This is because recent events may be surrounded with litigation while those more than about ten years old are difficult to establish since the memories of participants may begin to fail.

Details are obtained from both of the above routes and are combined into a knowledge base which may then be consulted by others. To do this, it is necessary to take the case histories, which are in the form of 'stories', and to transform them into a structured form suitable for manipulation by computer. The form chosen for this is the concept of an event sequence diagram.

4. EVENT SEQUENCE DIAGRAMS

Event sequence diagrams (ESDs) [13] show the temporal order and relationship of events leading up to a particular outcome. Fig. 2.1 [10] shows a typical diagram representing the collapse of a roof under snow loading.

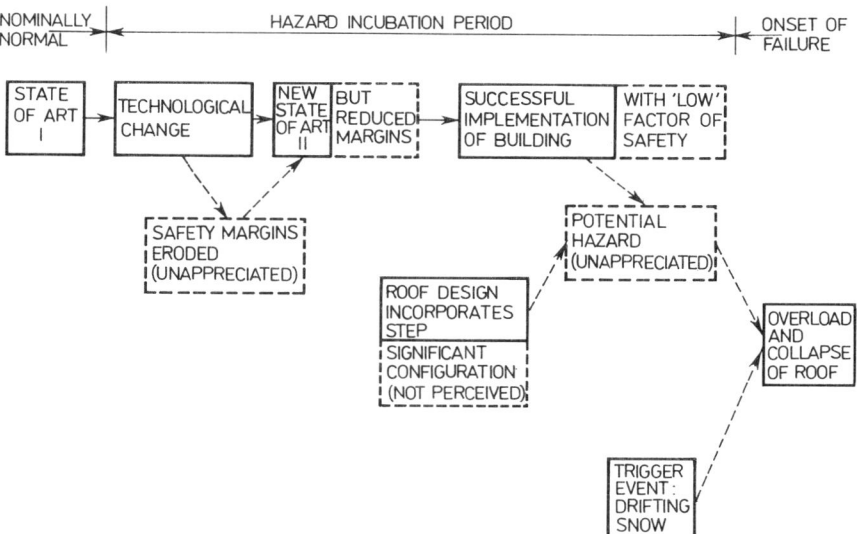

Fig. 2.1 — Typical ESD representing the collapse of a roof under snow loading.

To develop a useful knowledge base, it is necessary to collect a large number of case histories and to refine them into a series of ESDs. This raises the problem of defining a suitable 'vocabulary' for describing events. It will be shown that a 'learning' method depends on having propositional concepts ('words') which occur in more than one diagram, enabling a 'linkage' to be established. It is therefore proposed to establish an evolutionary 'dictionary' of concepts, sufficiently large to cover the richness of the range of cases held in the knowledge base yet small enough to ensure repeated use.

5. HIERARCHIES OF KNOWLEDGE

Fig. 2.2 shows a series of ESDs arranged hierarchically. The lowest (deepest) level contains the most detailed information, the stories of individual cas histories. At the other extreme, the highest (shallowest) level represents the accumulated 'story' of all the case histories in more general terms. This hierarchical structure is useful because it reflects the fact that in some situations very detailed information may be required while, in other cases general concepts are more meaningful. It enables an appropriate level of problem solving within the knowledge-based system (KBS). The use of a 'high level' concept, such as 'poor site supervision', may be sufficient for some purposes whereas a 'low level' concept such as 'reinforcement starter bars omitted', may be necessary for a more detailed analysis.

6. UNCERTAINTY

Two kinds of theoretical models can be distinguished. Following [14] a closed world model represents total knowledge about everything in a particular system and an open world model represents partial knowledge where some things are known to be true, some are known to be false and others simply unknown. Thus in a closed world every concept is either true or false and no undefined or inconsistent states are possible. In a closed world the information is complete in that all and only the relationships that can possibly hold among concepts are those implied by the given information. In an open world model there are four possible states of a concept, i.e. true, false, unknown and inconsistent, with intermediate stages in between. Most logicians like to forbid inconsistencies but in practical problems the finding and settling of inconsistency is an important element of the problem solving process.

A mathematical formalism is said to be complete if every formula which (in accordance with its intended interpretation) is provable within the formalism embodies a true proposition. Conversely, the formalism is complete if every true proposition is embodied in a provable formula. Unfortunately, Gödel has shown that this deepest level of mathematical completeness is unattainable in any formal system rich enough to contain arithmetic. However, in structural reliability theory this deep aspect of incompleteness need not be of concern in practical applications.

What is of concern, however, is the need to allow, in our theoretical framework, for the open world incompleteness. The central relationship is that between the measure of probability (support, belief, confidence, etc.) $p(A)$ for the truth or dependability of a proposition (event) A and the measure of probability for the truth or dependability of the negation of the proposition $p(\overline{A})$. In standard probability theory, if $p(A)$ is known, the theory forces the value of $p(\overline{A})$ to be $1-p(A)$; this can only be justified if the sample space from which A is taken is known or it is sufficiently accurate for all practical purposes to assume that it is accurately known. In any real engineering problem decisions have to be made under the incompleteness of partial ignorance since 'how can we know what we do not know?'.

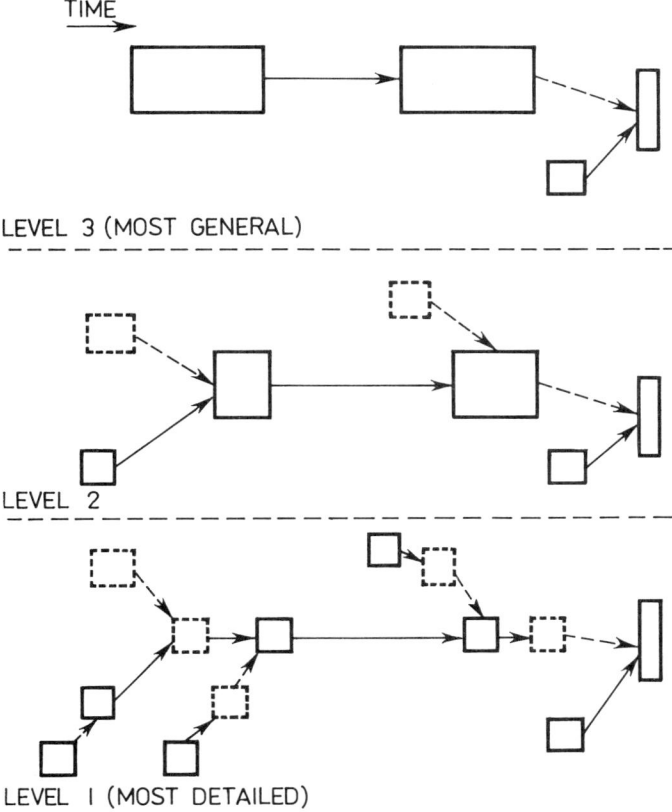

Fig. 2.2 — Hierarchical arrangement of ESDs.

Two recent theories of support logic [15] and interval probability [16] have been suggested. There are differences in the mathematical manipulations of these theories, but in essence they are very close, so that both can be discussed for the purpose of this chapter in terms of support logic.

In this theory two measures are associated with a concept (proposition or event): one, the necesssary support $N(A)$, is a measure of the truth or dependability of the concept and the other, the possibility $P(A)$, is a quite separate measure of the truth or dependability of the negation of the concept such that $N(A)=1-p(\overline{A})$. In general, $N(A)+N(\overline{A})\leq 1$. In this way, the measure of support for a concept is decoupled from the measure of support against the concept; the two measures are assessed quite separately. Three extreme states are therefore possible, i.e. support pair values of [0,0], [1,1], [0,1]. The first represents absolutely false since there is no evidence for and maximum evidence against. The second represents absolutely true since there is maximum evidence for and no evidence against. The third represents 'don't know' since there is no evidence for and no evidence against.

A measure of inconsistency can be obtained from support logic pairs since if two solutions are found for the support for a proposition A then any inconsistencies between those solutions can be found. For example, if we find A with support [0.3,0.6] and quite independently (implying the use of a multiplication model) we find A with support [0.4,0.7], then from Table 2.1

Table 2.1 — Determination of inconsistency

	$0.3 = N(A)$	$0.4 = N(\overline{A})$	$0.3 = N(A_u)$
$0.4 = N(A)$	$N(A \cap A) = 0.12$	$N(A \cap \overline{A}) = 0.16$	$N(A \cap A_u) = 0.12$
$0.3 = N(\overline{A})$	$N(\overline{A} \cap A) = 0.09$	$N(\overline{A} \cap \overline{A}) = 0.12$	$N(\overline{A} \cap A_u) = 0.09$
$0.3 = N(A_u)$	$N(A_u \cap A) = 0.09$	$N(A_u \cap \overline{A}) = 0.12$	$N(A_u \cap A_u) = 0.09$

Here $\overline{A}$ is the negation of A and A_u is the uncertain A such that $N(A)+N(\overline{A})+N(A_u)=1$. The support for A is $N(A)=N(A \cup A)=0.12+0.16+0.12+0.09+0.09=0.58$, and $P(A \cup A) = 1 - 0.12 = 0.88$. The conflict (A) is the support for $(A \cap \overline{A})$. Thus $N(A \cap \overline{A}) 0.16+0.09=0.25$ and $P(A \cap \overline{A})=1-N(A \cap A)-N(A \cap \overline{A})=1-0.12-0.12=0.76$.

the support for A is [0.58, 0.88] and the conflict $(A) = [N(A \cap A), P(A \cap A)] = [0.25, 0.76]$.

Fig. 2.3 is a Venn diagram for an open world set A with supports $N(A)$,

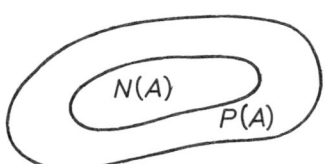

Fig. 2.3 — Venn diagram for an open world set A with supports $N(A)$, $P(A)$.

$P(A)$. Notice that the underlying assumption is that the sets A and $\overline{A}$ are crisp, but that the boundary between A and $\overline{A}$ is not known precisely. In this sense the spirit of a fuzzy set is captured. It would be possible to extend the definition of a probability interval $p(A) = [N(A), P(A)]$ to a fuzzy probability through the inclusion of a membership function over the interval so that $p(A) = \chi_{p(A)}(q)$, where $q \in [N(A), P(A)]$ and χ is the fuzzy membership function; however, this will not be pursued further for two reasons. Firstly, it introduces a level of complication to the algebra which is unwarranted. Secondly, and more deeply, the fuzziness inherent in any problem can be represented more effectively by the use of the concept of hierarchical modelling. Any decision of uncertainty modelling is based on choices made in the modelling of the phenomena under consideration and in particular the choice of sample space. If we think of all concepts as holons in a hierarchically structural knowledge base, then by looking upwards towards the infinite vague unity of the universe any concept (holon) is a part, and looking downwards to the precise infinitesimals of the universe any concept (holon) is a whole. The sample space represents a choice of holons which are convenient for solving a problem. In problems where the sample space consists for all practical purposes of mutually exclusive non-interactive concepts then classical probability is sufficient. However, where it is difficult to choose a sample space with these characteristics, then estimates of their interdependencies have to be obtained. In a sense fuzzy set theory is an attempt to change the level of definition (in fact to induce vagueness artificially) in order to enable better problem solving. Thus although the mathematical distinction between the membership function of fuzzy set theory and the probability function is clear and the conceptual distinction is valid, it is possible to model fuzziness without the use of fuzzy set theory. The algebra of interval probability or support logic could be applied to a fuzzy membership function or to a probability measure. The latter will involve one mapping from the power set to the interval [0,1] and the former will involve two mappings (the membership function from the sample space as well as the probability function from the power set) to [0,1]. The problem with this approach is that it can lead to an infinite regress since the membership function can itself be defined in terms of other concepts which also have membership functions.

For the purposes of practical computation, the interval probability of support logic as a measure on concepts (holons) arranged in a hierarchically structured knowledge base is a simple open world model of concepts which are more or less precisely defined. At high levels in a knowledge base concepts will be vague and therefore will tend to attract high levels of evidential support but at the expense of information content. At low levels the concepts will be precise and therefore of high information content by attracting consequently lower levels of support.,

7. MACHINE LEARNING

The topic of machine learning is an increasingly important area of artificial intelligence research, and the ability to learn from experience would be a useful ingredient of any meaningful KBS.

The method of 'learning' being developed is based around the use of an algorithm developed to aid clinicians in medical diagnosis [17]. Instead of observed symptoms and sets of possible diseases, concepts in ESDs and their outcomes are used. Learning is then carried out by induction — that is, general rules are proposed from the examination of a number of example cases. There are two phases to the 'learning' process involving algorithms for discrimination and connectivity which will now be described.

Fig. 2.4 shows three imaginary ESDs, in which concept names are represented by numbers. The ESDs are simple 'tree' structures, with each concept being a node of the tree. Each node may in turn be classed as a 'head' or 'tail' node when viewed from another node. For example, in ESD

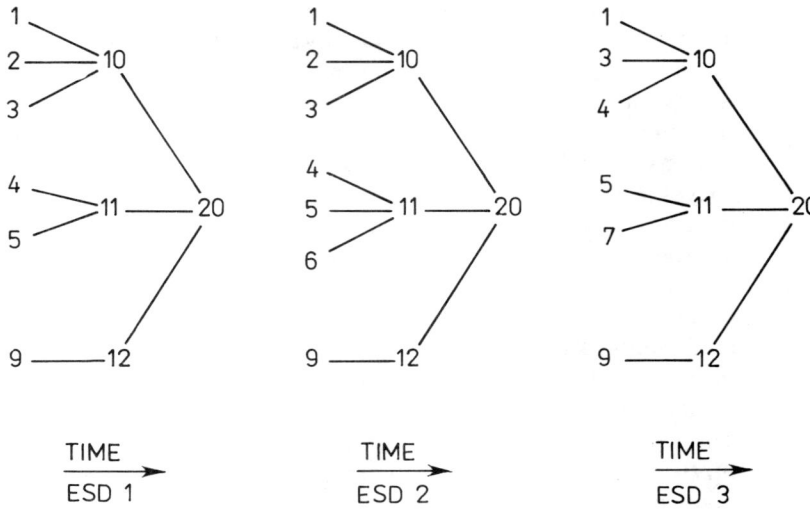

Fig. 2.4 — Three imaginary ESDs exhibiting 'tree' structures.

1, node 10 is a head node of nodes 1, 2 and 3 and a tail node of node 20. In Fig. 2.4 all three ESDs have the same final outcome, 20, and intermediate outcomes 10, 11 and 12. Thus, for example, node 20 might refer to 'failure', nodes, 10, 11 and 12 to 'human error', 'limit state' and 'random hazard' respectively, and nodes 1, 2 and 3 to 'wind loading (code) value found to be too low', 'no consideration of progressive collapse in design' and 'no consideration of explosive loading in design'.

The discrimination algorithm considers each head node in turn and calculates, from the evidence of all the ESDs, which tail node is most indicative of that outcome. For example in Fig. 2.4, node 9 occurs only as a tail node of node 12 and nowhere else. This suggests that in any other future case the observation of 9 is highly indicative that 12 may occur. Node 4 is a tail node of both 11, in cases 1 and 2, and 10 in case 3. The future occurrence of 4 is therefore evidence for probably 11 but possibly 12.

The connectivity algorithm adopts a parallel approach. Each head node is considered and a search is made for groups of tail nodes which commonly precede it. In Fig. 2.4 nodes 1 and 3 always occur together before node 10. The group of nodes (1, 2, 3) is also indicative of 10, but less strongly so since the three do not occur in each case.

For a more detailed discussion of these algorithms see Stone *et al.* [18].

8. A PRACTICAL EXAMPLE: THE SAFETY OF MINES

The legacy of old limestone mines in the Black Country of the UK, many of which were abandoned long ago, has resulted in sporadic incidents of subsidence over the past 100 years or more. Two types of event have occurred; first, local crown holes where a narrow shaft-like hole appears at the surface and, second, subsidence over a more general area of land. These incidents did not arouse undue concern until 1978 when subsidence in Wednesbury caused significant damage to warehouses and industrial plant. Consequently, the Department of the Environment in conjunction with the Metropolitan Borough Councils (MBCs) of Dudley, Sandwell and Walsall, together with the West Midlands County Council (no longer extant), commissioned Ove Arup and Partners to study the scale and nature of the problems, to make an assessment of the risks and to make recommendations for monitoring and remedial measures. The study was completed in 1983. It indicated that over the 300 km^2 studied, the total land surface area that might be disturbed is about 1.5 km^2. Indications of relative risk levels and possible treatments were given. Of the 100 recorded instances of surface disturbance, damage to property occurred on only 12 occasions over about 5 hectares with no loss of life. Thus the general risk is low. Limestone was worked by galleries along the strike and by room and pillar mining. The two seams worked, the Upper and Lower Wenlock Limestones, are 5 m and 10 m thick respectively. As a result, the mined cavities left after limestone mining were very large compared with those of coal mining; because of the strength of the limestone, the cavities remain open much longer. Thus further ground movements may occur. Consideration zones were defined within which consideration should be given to the need to investigate the ground in relation to surface movements that could be caused by deterioration of old limestone workings. These zones have subsequently been further subdivided into risk subzones according to the type of risk (crown holing or general subsidence), and into land use subzones, according to the type of land use.

Following this study, the Government allocated funds under the Derelict Land Grant (DLG) provisions to enable investigation, monitoring and remedial action to be undertaken to solve the potential problems due to old limestone mines. The Secretary of State for the Environment appointed the Black Country Limestone Panel 'to advise . . . on proposals for monitoring, site investigation and remedial works in relation to old limestone mines in the Black Country; and to consider other related issues as necessary.' As part of this work the panel makes recommendations to the Department of the Environment about applications from the MBCs for funding of treatment works under the DLG.

The objective of the work described [19] here was to produce a computer KBS to help the deliberations of the panel. The system is a model of data and expert opinion on the problem of the limestone mines in the West Midlands. It is emphasized that the system is an aid to the decision process, and therefore must not be viewed in any sense as a potential replacement for the human decision makers.

9. PERSPECTIVES

Solutions to complex problems are inevitably formulated within the perspectives of the key decision makers. The parties involved will have differing perspectives, attitudes and objectives and are therefore likely to have differing opinions about the best way to use limited resources.

Central to the limestone mine problem is the notion of risk and its role in the balancing of cost and benefit for the various proposals. There are obvious distinctions between local and national perspectives which may lead to differing judgements about the merits of various proposals, and the nature of the risks involved. There are, however, other perhaps less obvious distinctions. Professional advisers will tend to formulate problems from within the perspectives of their own education, training and experiences within their discipline. Thus one adviser may see a problem as straightforwardly technical and requiring an engineering solution whereas another adviser may formulate the problem in terms of economic, social or political criteria.

Any complex problem, such as that posed by any one of the Black Country limestone mines, has many aspects to it which each adviser may weigh differently. One of the purposes of the decision model presented in this chapter is to attempt to focus on these aspects and to provide a means for exploration and discussion of the weightings. Two obvious facets of all the limestone problems are the technical assessment of risk and the assessment of the quality of the usage of the land above the mines.

10. Knowledge classification

For the present purpose the knowledge and information available can be classified into three types which will present in various mixes in each aspect of a problem. These are facts, scientific models and descriptive models.

Facts are generally obvious, they are what is. However, facts are often established by measurement and analysis which contain hidden assumptions. For example, statistical data are susceptible to various interpretations. In the model, basic propositions were defined which constituted the facts of the problem; they were propositions which were considered to be detailed and precise enough for assessemnt by various experts for their degree of dependability. Each fact is known as a base relation and constitutes an 'atom' of the model of the problem. Associated with each fact therefore is the name of the assessor and a degree of uncertainty.

Conventional quantitative assessments of risk rely on scientific theories of capacity (R) and demand (S) and use probability theory to assess the chances of failure. However, a theory always has limitations which restrict the scope of its application and it is perhaps more useful to think of all scientific theories as models. It is, of course, the duty of an engineer to ensure that he is aware of these limitations both in deciding on a suitable theoretical model and in inerpreting the results from such a model. In the limestone problem there are no available scientific models of demand or capacity which could be used in a conventional risk analysis. Quantitative models such as the National Coal Board methods of predicting ground strain, given a collapse, do exist and can be used. However, they do not help in formulating the chance that a mine will collapse and of course they do relate to coal mines and not limestone mines.

Although causal scientific models do not exist, it is not true that the general mechanisms of general subsidence and crown hole formation are not understood. Linguistic descriptions of the mechanism of mine deterioration leading to roof falls, possible pillar collapses and subsequent subsidence or crown hole events are more or less agreed among experts in geology, mining and civil engineering. The model of the technical risk described in this chapter is an attempt to capture that agreed account. Those experts who were consulted and cooperated in the process of producing the model generally found that the model has largely represented their understanding. The risk analysis is therefore a heuristic one based on a descriptive account. There is no attempt to predict statistical chances of subsidence or crown hole events, but rather to rank the severity of the risks from the various mines to help in establishing priorities for treatment strategies.

By contrast, the problems associated with obtaining assessments of present and future land usage were much more difficult. The experts in this area of the problem (development economists and planners) were able to provide full descriptive accounts, but they did not feel able to identify the cocneptual structure of the problem and the quantitative relationships between those concepts to participate in the construction of the model. The authors therefore based the model on their own interpretation of the descriptive account.

11. THE MODEL

The model consists of a set of propositions. The propositions are organized into a hierarchy according to their level of definition, and they are connected by logical relations. These two aspects of the model will now be described.

11.1 Hierarchical modelling

The propositions which make up the model are ordered according to their precision of definition. At the top of the hierarchy are the propositions which are the most vague and imprecise; at the bottom are those which are as precisely defined as is required for the problem, the base relations or facts as described earlier. The technique for the development of the hierarchical model is conceptually straightforward, but requires some practice. This first task is to make a statement about what is required in as vague a way as is necessary to ensure there is little or no dispute about it. Thus, for example, a top level statement for the decision analysis concerning the quality of a structural design might be of the form 'The design is good'. Of course everyone is likely to agree with the proposition that a design should be good because it is vague and of little practical use. The next step is, therefore, to write down propositions which together are logically sufficient for the truth or dependability of the first statement at a level of detail which is only slightly more precise. A very basic question now arises: how can the sufficiency of these propositions be ensured for the first proposition? The answer is of course that it cannot.

However, in practical terms, as many of the necessary conditions can be listed as can be thought of and the basic incompleteness of this or any other model recognized. Thus for the structural design example the top level proposition may be developed as 'The design is good if the design is safe and the design is economic.' Here the quality of design has been modelled by two subconcepts, safety and economy, which are in turn vague and imprecise. However, other concepts such as functional requirements and environmental impact have not been included and various views could be held by the experts and likely users of the model as to their importance or relevance. Clearly for completeness they should be included but, if there is disagreement, the form of the rule provides a basis for discussion.

The process of writing down lists of necessary conditions for the truth or dependability of higher level propositions in the hierarchy is continued until the level of definition of the proposition is precise enough for the purposes of the problem being modelled. These then are the base relations.

The formulation of the part of the model concerning technical risk was a three-stage process. First, experts in geology, mining and civil engineering were interviewed so that the authors could gain an understanding of the issues involved. Second, a draft version of the hierarchical model was written and, third, this model was critically reviewed by the experts consulted. In the event little alteration to the draft was required. The formulation of that part of the model concerned with land usage and the general strategy of the panels decision making was carried out by the authors on the basis of specialist written reports made available by the Department of the Environment.

As explained earlier, no attempt has been made in the model to estimate the chances of occurrence of a general subsidence or crown hole event. Instead the relative likelihood of these events from the various mines has been estimated by a simple ordering. Similarly the damage to property or threat to loss of life or potential land use has only been estimated in relative terms. The benefits of various strategies have been treated similarly.

11.2 Logical relations and evidential support

Each statement in the development of the hierarchy is a rule of the form

$$A \text{ if } (B_1 * B_2 * B_3 \ldots B_n)$$

where (A, B) are propositions, (if, *) are logical relations and * represents (and, or). The negation of a proposition B was also used (not B). As mentioned earlier, the propositions $(B_1 * B_2 * B_3 \ldots B_n)$ are considered to be logically sufficient for A, i.e. $(B_1 * B_2 * B_3 \ldots B_n) \supset A$, where $\supset$ is the implication logic operator.

Associated with each proposition is a series of attributes. For the structural design example previously mentioned these may be, for example, (p, q) in 'the design is safe (p, q)', where p is the name of the design and q is a safety factor value. Associated with each set of attribute values for a proposition is a truth or dependability measure. In the model described here, a new system of support logic was used in which two measure on a scale (0,1) were adopted. The first measure is called the necessary support (N) for the proposition. It is a measure of belief in the maximum dependability that the set of valuations of the attributes surely belongs to the concept represented by the proposition. Thus for example 'the design is safe (design_no_1, 1.6)' with a necessary support of 0.4 means that it is believed to the level 0.4 on a scale (0,1) that the design_no_1 with a safety factor of 1.6 is certainly a safe design. The second measure is called the possible support (P) for a proposition. It is related to the necessary support for the negation of the proposition. It includes those elements of belief which are not sure (i.e. not necessary) but are possible. In the support logic numerical scheme, $1-P$ is associated with the necessary support for the negation of the proposition and $N+(1-P) \leq 1$ or $P \geq N$. This is because an individual is inconsistent if he believes something is surely the case and is surely not the case at the same time. Thus the maximum values of the necessity for the case (N) and the necessity against the case ($1-P$) must sum to less than 1.

The method of representing the logical rules was the modified Horn clause form of PROLOG

$$A:B_1, B_2, \ldots, B_n:[N,P]$$

where $A, B_1, B_2, \ldots, B_n$ are atoms, and A is known as the head of the clause and $B_1, B_2, \ldots, B_n$ as the body of the clause. It is a PROLOG clause with the addition of the support pair $[N, P]$. For each assignment of each variable, if $B_1, B_2, \ldots, B_n$ are all true, A is necessarily supported to degree N and $\overline{A}$ is necessarily supported to degree $1-P$.

If the body of a clause is a compound proposition, it will be necessary to determine the support pair for it from the support pairs for its atoms. For example, if $A:[N_1, P_1]$ and $B:[N_2, P_2]$ are given, how are $(A \text{ and } B):[N_3, P_3]$ or $(A \text{ or } B):[N_4, P_4]$ determined? This chapter does not describe the details of the support logic calculations which have been described previously. In summary it has been shown that the operations of fuzzy logic and probability theory are point estimates on an interval of possibilities which are, for example,

$$\max(N_1+N_2-1,0) \leq N_3 \leq \min(N_1, N_2) \quad \text{(necessary support for } A \text{ and } B\text{)}$$

$$\max(P_1, P_2) \leq P_4 \leq \min(1, P_1+P_2) \quad \text{(possible support for } A \text{ or } B\text{)}$$

In any particular application, inferences made carrying through the full possible interval between the maxima and minima defined above would soon degenerate to fill the full interval (0,1) and be of no practical use. Two particular models for assigning unique support values have been suggested. These are the multiplication model where

$$N_3 = N_1 N_2 \qquad P_3 = P_1 P_2 ,$$

and the minimum model where

$$N_3 = \min(N_1, N_2) \qquad P_3 = \min(P_1, P_2)$$

In both cases

$$N_4 = N_1 + N_2 - N_3 \qquad P_4 = P_1 + P_2 - P_3 .$$

The relationships between these operations and those of probability theory, fuzzy logic and the Dempster–Shafer theory of evidence have been fully discussed.

12. KNOWLEDGE BASE

As outlined earlier, the knowledge base consists of a set of hierarchically organized and logically related propositions. The hierarchy was established by starting with the general proposition 'Approve scheme if benefit positive and costs acceptable'. The lists of necessary conditions were developed as described in the preceding section to the level of detail shown in [19]. It is emphasized again that this decision scheme is not the scheme used by the panel to arrive at its decisions but was produced as an aid to that process. For example, one of the rules reads in English as 'Benefit positive if the scheme prevents serious consequences or the scheme allows new opportunities in the subregional context or the scheme provides necessary information to take a responsible decision or the public is reassured.'

In the language of the support logic knowledge base this same rule is

bene_pos (A, B):—prev_con (A, B) or new_opp (A, B) or

nec_info (A, B) or pub_reass
(A, B):[1.0, 1.0]

where:— is the symbol for if, and (A, B) are the attributes (scheme name, consideration zone name) of the propositions or relations with which they are associated. All of the rules were allocated initial importance values of [1.0,1.0].

Inputs to the knowledge base are subjective estimates (expressed as a support pair (N, P)) of the truth or dependability of all the base relations (facts) for a given mine, treatment scheme or land usage. Associated with each input value is the name of the assessor. For example, the consulting engineers investigating the condition and extent of the mines assessed for each mine the dependability of the statement 'There are significant faults normal to bedding which intersect strata above the mine roof'. In the knowledge base this opinion is associated with the base relation:

> sig_fault (scheme_number, consideration_zone_name,
>   mine_name,
>   risk_sub_zone_name, treatment_method,
>   assessor_name)

Similar assessments were made for all the other base relations concerning physical risk. Inputs to base relations concerning land use were derived by the authors from descriptive accounts and assessments made by land use consultants. Judgements expressed by representatives of the MBCs were similarly interpreted and included where possible, although unfortunately not in every case.

The knowledge base for the physical risk was tested in two ways. First, the hierarchy of rules was examined and accepted by all the experts consulted before inputting them into the computer. Second, the computer was interrogated to ensure that the answers were as expected for sample queries. This was done to check the logic calculations and to check that the input judgements led to intuitively reasonable answers and explanations. The knowledge base for the land use was checked to ensure that the answers at least complied with the assessments made by the land use consultants.

13. INTERPRETATION OF MEASUREMENTS

Data collected from observations and measurements on full-scale civil engineering systems are often difficult to interpret because of the inherent uncertainties involved in the behaviour of such systems. Measurements arise broadly from two processes, namely monitoring and testing. The behaviour of a civil engineering system may be monitored by measuring loadings and movements of large structures such as dams and offshore structures or by measuring rainfall and streamflows in hydrology or ground movements and pore water pressures in geotechnics. Systems may be tested in the quality control of materials, in establishing structural integrity or in the investigation of a particular site.

Monitoring is so obviously essential to gain better understanding of the behaviour of a system that it is perhaps rather surprising that it is not commonplace. One of the reasons is that the interpretation of the results is not easy. One example of where extensive monitoring work has been carried out is by the Italian power generating authority ENEL in collaboration with ISMES [20, 21]. Automatic monitoring systems for several concrete dams in Italy have been implemented. With the development in the last few years of modern instrumentation for automatic recording and telemetry of digital data and microcomputers for data processing, the difficulty and cost of gathering and storing data have been greatly reduced.

The abundance of data raises its own problems in terms of evaluation and interpretation. Firstly, the availability and cost of the expertise required to interpret the data limit the extent and effectiveness of monitoring. Secondly, the volume of data can be such that it cannot be handled easily and the vital task of recognizing potentially important information in a mass of routine information becomes very difficult.

14. THE REQUIREMENTS

A KBS for use in the monitoring of an engineering system [22] should be able:

(i) to recognize and represent the important features of the data;
(ii) to compress the volume of data without missing important features;
(iii) to represent important features at a level which will allow reasoning with and about them;
(iv) to make inferences based on the data representation concerning the state of the system.

Again it must be stressed that such a KBS can only advise and assist in an interpretation of the data and that the final interpretation, decision and responsibility will always rest with the engineer.

The essential purpose of monitoring is to enable a decision to be made about what action (if any) is necessary given the current state of the system. There are a number of inferences the engineer may wish to make from the data, for example:

(i) to determine the current state of the system;
(ii) to predict possible and probable future states of the system;
(iii) to determine the reasons for the current state of the system;
(iv) to determine whether the monitoring system is functioning properly;
(v) to determine what change in the structure of the system is implied by the data.

15. SIGNAL ANALYSIS

Signal analysis in civil engineering tends to be dominated by the use of time series analysis based on classical statistics. These techniques have proved extremely useful in the analysis of long time histories, but are difficult to relate to human interpretation of data. Estimates of means, correlation coefficients, spectral coefficients, etc., are useful in algebraic analysis but they represent a drastic and sudden compression of the information in the data. These measures carry little information about the shape of signals (except in the case of periodicity), particularly local shape features which seem to be of particular importance in human interpretation. Frequency analysis has been useful in providing transformations which reveal features otherwise hidden in the data, but again their interpretation requires careful consideration of the shape of the results.

The features which can be recognized in signals can be classified according to two criteria, shape and value, which apply in both a global and a local sense. For example, a signal has global shape properties, such as smoothness, variability and periodicity, and local shape properties, such as peaks, troughs, slopes and other regularities. The global value properties are overall mean and trend whereas the local value properties will be maxima, minima, extreme values and local means.

The signal is, of course, a partial representation of the system being monitored. It is a collection of measurements through time which are a mapping from the actual real system to the signal dimensions and as such the signal is a representation constrained by the observability of the system, its state variables and the equipment available. Any KBS written to interpret patterns in data must address the nature of these constraints and the uncertainties they create in the relationship between the signal and the actual system.

The basis of the proposed methodologies to represent the shape of a signal is that a signal can be modelled as a sequence of vectors. The vectors used can be drawn from a limited set of possible types. The number of vector types is specified initially and the range of possible slopes ($-90°$ to $+90°$) is divided into this number of classes. Each class has a class vector which has a slope at the midpoint of the class, and a slope class symbol. The vectors are all of the same initially specified length.

A computer program SHAPQUAN has been written in FORTRAN to control the process of compressing the shape information contained in the signals and to enable the building of a granularity hierarchy of shape representation. This has been reported [22].

Syntactic pattern recognition involves the identification of the structure of patterns through a hierarchical composition of simpler subpatterns. The simplest subpatterns are known as pattern primitives. The way the patterns are combined to form higher level patterns is known as the pattern grammar. Each pattern or structure to be recognized has its own grammar. A pattern is represented at the lowest most detailed level by a sequence of pattern primitives called a sentence. A sample pattern belongs to a language if the sentence contains pattern primitives of that language and a syntax analysis or parsing of the sentence is correct within the pattern grammar.

The principles of syntactic pattern recognition have been used for example in medicine to recognize features in signals resulting from monitoring body functions [23]. In this work a computer program SHAPE has been written in PROLOG to define a pattern grammar with pattern primitives which are the chain codes resulting from the program SHAPQUAN, mentioned earlier.

The shape representation system described above will provide the same representation for signals of the same shape but different values. The value representation system allows a controlled compression of information about the value of a signal. The system comprises two parts. A program VALQUANT (written in FORTRAN) is a dynamic granularity hierarchy which generates a particular level of representation on demand. A program VALUE, written in PROLOG, contains a type hierarchy for concepts of value. The basis of these techniques has been reported and used for the nondestructive testing of a pile by the 'shock' method. In this method the movement of the head of a pile is monitored after it has been struck by a hammer [22].

16. USE IN KBS

The data representations described can be used in at least two related ways. Firstly, they can be processed in some way to yield further information or they can be used for reasoning and inference. The chain code representations can be integrated, differentiated, autocorrelated and cross-correlated with other chain codes. The compressed representation of the data given by the chain code reduces considerably the amount of processing.

Using the representations derived from these methods it is possible to reason with them within a KBS. For example,

(1) IF   signal_a IS extremely_low AND
        signal_b IS high OR extrmely_high
  THEN dam is unsafe.
(2) IF   signal_a HAS sharp_peak AND
        signal_ HAS deep_trough
  THEN equipment_malfunction.
(3) IF   similarity (signal_a, signal_n) IS<0.6
  THEN represent signals as chain codes and recalculate
(4) IF   time_difference (signal_a, signal_b) is large_lag
  THEN delay IS excessive.

The pattern-directed inferencing of a system such as PROLOG means that it is not difficult to represent patterns of parameter states which represent some important status of the system. and then to use 'demons' to look for these in incoming signal representations.

To obtain greater insight into the information contained within a signal, it is often useful to transform time series into the frequency domain. The shape and value representation system can then be used to characterize these data in a similar way in order to obtain high level descriptions of periodicity, randomness and noise.

17. CONCLUSIONS

(1) A fundamental question is 'Should society expect no failures?' The answer must be 'no' because such perfection cannot be reached and measures of risk are, by their nature, problematical.
(2) It is argued that engineering is a process of problem solving and decision making using information with varying degrees of dependability. Dependability is necessary but not sufficient for truth.
(3) The notion of the inductive reliability of a hypothesis should be replaced by the notion of a responsibility to act on that hypothesis.
(4) Measure of risk are not absolute measures but are aids in the process of the management of knowledge to control risk.

(5) 'Expert systems' would be better described as 'advice systems' because the computer is only able to offer advice and cannot replace the engineer as a responsible decision maker.
(6) Work towards the development of a KBS which might be an aid in the management of the safety of a project has been described.
(7) Case histories of engineering projects are being collected in the form of ESDs. The knowledge base consists of hierarchical structured set of concepts and relations.
(8) An open world mathematics of uncertainty has been discussed and the concept of evidential support introduced.
(9) A method a 'machine learning' has been briefly described which involves two phases with algorithms for discrimination and connectivity analysis.
(10) The development of a KBS for risk and cost benefit analysis of limestone mines in the West Midlands of the UK has been briefly described.
(11) Measurements taken from full-scale monitoring and testing of civil engineering systems are often difficult to interpret because of the inherent uncertainties. The requirements of a KBS to assist in this have been outlined, and computer program briefly mentioned.

REFERENCES

[1] Alty, J. L. and Coombs, M. J., *Expert Systems*, NCC Publications, Manchester, 1984.
[2] Forsyth, R., *Expert Systems*, Chapman and Hall, London, 1984.
[3] Blockley, D. I., Pilsworth, B. W. and Baldwin, J. F. Structural safety is inferred from a fuzzy relational knowledge base, *ICASP-4, Florence, 1983*.
[4] Fenves, S. J., Maher, K. L. and Stiram, D., Expert system: CE potential, *Civ. Eng. (ASCE)*, **54** (10), 44–47, 1984.
[5] Ogawa, H., Fu, K. S. and Yao, J. T. P., Speril II — an expert system for damage assessment of existing structures, *Technical Report CE-STR-84-11*, Purdue University, 1984.
[6] Duda, R. O., Gaschnig, J. and Hart, P. E., Model design in the PROSPECTOR consultant program for mineral exploration. In Michie, D. (ed.) *Expert Systems in the Microelectronic Age*, Edinburgh University Press, Edinburgh, 1979.
[7] Shortliffe, E. H., *Computer Based Medical Consultations: MYCIN*, Elsevier, New York, 1976.
[8] Blockley, D. I., *The Nature of Structural Design and Safety*, Ellis Horwood, Chichester, 1980.
[9] Settle, T., Scientists: priests of pseudo-certainty or prophets of enquiry?, *Sci. Forum*, **2**(3), 21–24, 1969.
[10] Pidgeon, N. F., Blockley, D. I. and Turner, B. A., Design practice and snow loading — lessons from a roof collapse, *Struct. Eng.*, **64A** (3), 67–71, 1986.

[11] Pidgeon, N. F., Blockley, D. I. and Turner, B. A., Site investigations — lessons from a late discovery of hazardous waste, *Struct. Eng.*, October 1988.
[12] Blockley, D. I., An AI tool in structural safety control. In: Nowak, A. S. (ed.), *Modelling Human Error in Structural Design*.
[13] Turner, B. A., *Man Made Disasters*, Wykeham, London, 1978.
[14] Sowa, J. F., *Conceptual Structures*, Addison-Wesley, Reading, MA, 1974.
[15] Blockley, D. I. and Baldwin, J. F., Uncertain inference in knowledge-based systems, *Proc. Am. Soc. Civ. Eng., Eng. Mech.*, April 1987.
[16] Cui, W. C. and Blockley, D. I., Interval probability theory for evidential support, submitted to *Int. J. Intell. Syst.*
[17] Norris, D., Pilsworth, B. W. and Baldwin, J. F., Capital diagnosis from patient records, *Fuzzy Sets Syst.*, **23**, 73–87, 1987.
[18] Stone, J., Blockley, D. I. and Pilsworth, B. W., Towards machine learning from case histories, submitted to *Civ. Eng. Syst.*
[19] Blockley, D. I. and Henderson, J. R., Knowledge base for risk and cost benefit analysis of limestone mines in the West Midlands, *Proc. Inst. Civ. Eng., Part 1*, **84**, 539–564, June 1988.
[20] Fanelli, M., Automatic observation and instantaneous safety control of dams: an approach to the problem, *Internal Rep.* ENEL/CRIS, Milan, 1979.
[21] Anesa, F. and Bonaldi, P., On line monitoring of dams during their operation: Italian Experience, *Proc. Conf. in On-Line Surveillance and Monitoring, Venice, May 1986*, pp. 62–83.
[22] Comerford, J. B., Blockley, D. I. and Davis, J. P., The interpretation of measurements from civil engineering systems, submitted to *Civ. Eng. Syst.*
[23] Skordalakis, E., Syntactic ECG processing: a review, *Pattern Recognit.*, **19** (4), 305–313.

3

Structural assessment in a combined symbolic–numeric environment

T. Krauthammer,
University of Minnesota

1. INTRODUCTION AND BACKGROUND

Structural damage assessment following a major disaster, such as a flood or an explosive event, could be one of the most difficult tasks for an engineer who is required to inspect the structure, to determine its condition and the reserve capacity, and to explain the sequence of behavioural events that caused the observed condition. These requirements can be achieved by employing an expert for performing the inspection; this, however, may not be always possible. Furthermore, even an expert may need a considerable amount of time in order to assess carefully a structural system, and in the event that the assessment is required in a very short time the expert may not be able to perform it. These conditions create an excellent environment for the development of an 'expert system' which can represent the knowledge of an expert but perform the assessment in a much shorter time. An expert systemm for the post-event assessment of reinforced concrete structures subjected to explosive loading was presented previously by Krauthammer and Kohler [17], where a limited expert system was developed and employed for performing simpler tasks, and it was used to evaluate the required procedures. Similar efforts by Carson *et al.* [6] and by Ross *et al.* [23] were aimed at addressing the same problem. More advanced approaches to the problem were developed later by Krauthammer [18,19], by Ross [24], and by Wong *et al.* [28]. As far as the assessment of concrete dams, the publications by Franck and Krauthammer [10–12] represent the latest approach to that problem.

The development of an expert system is founded on the principles of artificial intelligence combined with an extensive knowledge base on the particular subject. The main contribution of artificial intelligence is the modelling process required for progression through a collection of rules, by making use of heuristic search, as discussed by Buchanan and Shortliffe [5], by Hayes-Roth *et al.* [14] and by Rich [22]. The process can be performed by starting from a set of initial states and progressing toward the goal state (forward chaining), by consideration of initial conditions and preconditions that allow sensible decisions to be made at each intermediate step. The progression to the goal state can be executed also in a backward manner (backward chaining), post conditions are set so that the progression from the goal state back down to the initial states meets all the conditions of all the intermediate states in a logical way, until the preconditions of the actions necessary to initiate the chain are met. The system can then devise a plan of action to achieve the goal by reversing the sequence of actions found during the search. The purpose of an expert system is to provide a search procedure by making use of plausible inferences based on knowledge supplied by the user, and the possible inferences are stored as rules in the program. The three major components of an expert system are as follows: the knowledge base that contains the general knowledge on how to solve the particular problems; the context or global database that describes the state of the dynamic solution, which is continuously updated; and the inference mechanisms that manipulate the context using the knowledge base.

Expert system for the task of structural evaluation usually contain the following four types of knowledge:

(1) general background knowledge in all the fields that are related to the engineering system under consideration;
(2) reliability knowledge that covers the ability to reason, judge, and develop well-defined reasoning about the failure mechanisms of an engineering system;
(3) statistics knowledge for applying statistical models to the evaluation of behavioural and failure uncertainties;
(4) knowledge engineering for capturing and reprsenting the first three types of knowledge in the expert system.

Each of these types of knowledge can be divided into two subgroups. The first is defined as 'shallow' knowledge that is used by an expert for solving specific problems, while the second is the 'deep' knowledge that can be used by an intelligent individual (who may not be an expert in the field) for solving the same problem, as discussed by Waterman [27]. It is clear that shallow knowledge is a result of accumulated experience from continued activity in a specific field. Deep knowledge, however, represents a fundamental understanding of an entire field, and is employed for solving problems that have not been previously encountered. Furthermore, shallow knowledge is contained in deep knowledge, and an efficient expert system should contain both types of knowledge. The shallow knowledge will be organized in special modules for performing specific operations, and these modules will be activated by rules that are formulated according to the deep knowledge. The shallow knowledge can be obtained by capturing the actual experience of experts or by employing information presented in the literature. These two approaches were implemented in the expert systems that will be discussed later herein.

The potential of expert systems in the field of construction and structural engineering was presented by Adeli [1], who combined chapters written by leaders in the field. It is possible to learn the reasoning process of an expert, to perform the work much faster than the human expert, and to expand the modelling and creativity aspects of structural engineering well beyond the potential reached by the deterministic analysis capabilities that have been so intensively developed in the past. Adeli presented also an excellent review and summary on the state of the art in expert systems development, and on the potential applications of expert systems in various engineering fields.

As far as the present problem is concerned (i.e. structural assessment following severe load applications), the expert systems are required to perform an assessment of a given structure and to provide clear decisions with respect to its condition and future use. These requirements are very similar to the diagnostic procedures performed by a physician in order to determine the health status of a patient and, on the basis of the findings, to devise a plan of action in order to treat the diagnosed problems. The topics discussed by Buchanan and Shortliffe [5] were aimed at this particular application of an expert system, and it was demonstrated very effectively that the diagnostic process can be performed by incorporating a medical knowledge base into the structure of an expert system. Based on those findings it is clear from the outset that the concept of a structural diagnostic expert system can be developed in a similar manner, and the main issues to be considered are the difficult problems of knowledge acquisition, knowledge representation, and inference mechanisms. Furthermore, in the diagnostic process, the expert is provided with existing symptoms and is required to define the causes that led to 'the current state of the world'; hence a combination of forward and backward chaining approaches is required for simulating this process.

For practical expert system development, one can employ an 'expert system development tool' which contains the framework for an expert system, and the user has to provide the knowledge base in the form of 'rules' that would be incorporated into the system. The expert system development tool selected for the two expert systems to be discussed later herein had to meet the following requirements:

(a) forward and backward inference and explanation capabilities;
(b) knowledge representation by frames and rules, and graphic displays;
(c) application of meta-knowledge for directing the search process;
(d) introduction of user-defined functions, or subprograms, for performing calculations within the knowledge base;
(e) the option to write AI or other routines in the development level;
(f) the flexibility to combine symbolic and numeric evaluation approaches;
(g) development and application on microcomputers.

Also, once the system is created, it would be possible to employ it for tracking the logic flow in it, and to study how changes in the behavioural aspects of the structures (e.g. changes of rules, or varying their location in the inference network) would affect the final decisions. The understanding of these issues would lead to the creation of effective inference mechanisms that would reduce the amount of search through the knowledge base for locating specific facts. Furthermore, in many cases one may not be able to provide all the required information by observing the damaged structure. This could be a result of not having access to various parts of the facility and/or not having time to perform all the required measurements. As a result, it would be necessary to couple the expert system with numerical codes which can be used to supplement the available information and to verify conclusions reached by the expert system. Such coupling of symbolic and numeric processes creates a more effective structural assessmnt capability, and this topic is the main issue addressed herein.

There are expert systems which combine symbolic and numeric approaches for performing specific tasks, but the numeric part is for representing approximate reasoning by means of probabilistic computations. This approach has been adopted because of inherent uncertainties in some of the answers provided by the user, as discussed by Ross [24], Blockley [4], and others. These numerical evaluations of the parameters within the expert system are not the type of calculations with which one is concerned when a 'combined symbolic–numeric' approach is developed. The present emphasis is on the coupling between AI processes and the mathematical solution of physical problems for reaching meaningful decisions about an engineering problem, as will be discussed further.

The expert systems to be discussed here were developed along the lines presented in the book by Adeli [1] since the technical issues under consideration seem to fall in the range of applications for which expert systems could be most suited, namely, well-defined engineering problems that are controlled by a relatively small number of (complicated) parameters. The solution of such problems requires the attention and knowledge of experts who, in many cases, may not be available on site or may not be able to address the problem in the short time that would be provided. Nevertheless, the expert system has to provide clear recommendations on the conditions of the structure, and whether its condition could endanger the lives and property of individuals located inside (or in the structure's vicinity). Because of the possibility of imminent failure (either with or without the application of additional loads), these decisions have to be made in a very short time, and all means have to be mobilized to facilitate the required outcome.

2. COMBINED SYMBOLIC–NUMERIC EVALUATION

The need to combine symbolic and numeric approaches is not new. Almost all solutions of physical problems contain some form of such a combination, as illustrated in Fig. 3.1. The simplest example for such a combination of approaches is the same diagnostic session at the physician's office. In most cases, unless the situation is very obvious, the physician will ask questions, but then will require the patient to undergo some tests. The types of tests to be performed will depend on the initial clues obtained from the answers provided by the patient, but the results from the tests will be coupled with other information for obtaining (or confirming) a diagnosis and the corresponding treatment. A similar combination of symbolic and numeric approaches has to be developed for addressing the present problem.

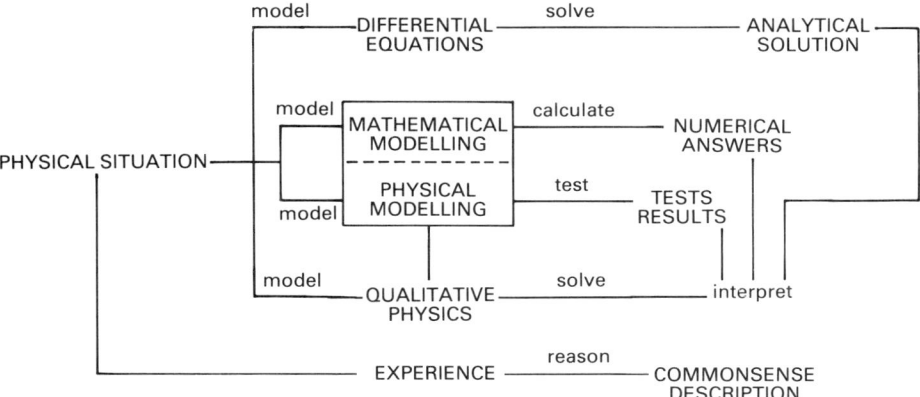

Fig. 3.1 — Combination of symbolic and numeric approaches.

Artificial intelligence languages, such as PROLOG and LISP, were developed for operations based on symbolic arguments, while 'traditional' programming languages, such as FORTRAN, BASIC, Pascal, etc., are aimed at processing mathematical equations. When symbolic languages are used for scientific computations, there is a significant reduction in the processing speed, while the scientific languages are not efficient for string operations. There are several ways for coupling symbolic and numeric programs, and each approach may have different advantages or disadvantages. The first approach for such coupling is to have the two different types of programs integrated into a single unit in the form of separate subroutines, or by having results from one part of the program stored as disk files to be called by the other part of the same program. This approach requires extremely careful programming in order to have the two parts interact satisfactorily, and in the event that changes are introduced into one part of the program corresponding changes may be required in the other part. Therefore, this approach is quite suitable for progress which may not require major modifications over time, such as in the expert system described by Franck and Krauthammer [9,11] but it will be too costly for programs that are updated frequently as a result of new information obtained by continuous research, as represented by the previous expert systems developed by Krauthammer [18,19]. Furthermore, single-unit coupling is very rigid, and any modifications may require a complete rewriting of the program.

Another approach is to create different modules for the numeric and symbolic parts that are actually independent programs, but which can be linked through another program acting as the 'controller' of the combined symbolic–numeric process. The issue of coupling these two distinct programming capabilities has been discussed by Degroff [8] and by Wong et al. [28]. According to these sources, there are five requirements for developing a successful coupled symbolic–numeric program, as follows:

(1) an ability to call external programs, and to switch processing modes;
(2) an ability to control data flow into the called program modules;
(3) an effective control of location of modules, and the timing of switches;
(4) an ability of one module to access information generated by another;
(5) an ability to trace and explain the flow of actions and values.

Items (1), (2), and (4) depend on the operational capabilities of the expert system (or the expert system development tool employed for creating the expert system), item (5) is an integral part of the knowledge base, and item (3) is related to both the knowledge base and to the inference networking.

The interfaces between program parts can be achieved by one of the following five approaches:

(1) The simplest approach is the use of batch files, where programs are executed sequentially and data are transferred from one part to the next through such files. This approach usually requires many read and write operations which tend to slow the independently efficient programs. Furthermore, one may have to make several formatting changes for such data in order to fit the various subprograms.
(2) The use of command lines which pass parameters, request the execution of a specific program, and provide all the information which is required for that operation.
(3) Different programs run on different computers, and one has to create the necessary communication.
(4) A 'master program' (or 'linker') controls the various need for names, data types, and addresses by creating procedure level programs.
(5) By employing a controlled access environment, which includes screen memory areas for text transfer, windowing for multiprocessing capabilities, and the implementation of memory block reservations and blackboards. This approach includes a 'director' that controls and supervises access to the various parts of the coupled program and the allocated storage.

Several expert systems which were developed recently employ different approaches, as follows. The expert system developed for the field assessment of concrete dams [9–11] is based on the first approach for creating a coupled program. The symbolic and numeric parts are integrated into a single program, and the numeric calculations are performed only if the user wishes to perform them (and, of course, if there are data for performing calculations). The expert system for protective structures applications [28] is composed of independent modules that can be accessed through a central program. Furthermore, one module can access other modules through this approach. The coupling is performed by procedures based on items (1) and (2) above. This type of approach has the flexibility required for a combined symbolic–numeric environment, but is quite simple for implementation. The expert systems developed by Carson *et al.* [6], Krauthammer and Kohler [17], Krauthammer [19], and Ross *et al.* [23] do not contain coupling between symbolic and numeric components (the use of probabilistic methods for approximate reasoning has not been considered as a numeric approach). However, a new expert system that is currently under development by this author is a coupled system which combines the capabilities of the previous expert system [19,20] with numerical options developed separately (for example [18,20]) and design options [21]. In the present effort, the approach is based on items (1)–(3) mentioned above. The two expert systems, one for the analysis of concrete dams and the other currently under development, are discussed in more detail in the following.

3. ENGINEERING EVALUATION FOR HAZARD PREVENTION

The problem of assessing damaged structures and rapidly reaching conclusions about their safety is, of course, complicated. The engineer must perform some measurements; then he needs to obtain some numerical results about the system under consideration, and only at that point may he be able to reach a decision about the future of the structure. This is a rather slow process, which may not be adequate under emergency conditions. First, the issue of gathering information about the structure, a very time consuming effort on its own, has to be reduced to a minimum. This can be accomplished by having all the engineering information about the structure in a computer 'memory bank', and this may be on a different computer located elsewhere. Each structure will be identified by a specific code, and that information can be retrieved from storage in the correct format for the expert system. The information that would be required at that time could be limited to observations on the conditions of the facility, and possibly data from some simple measurements. From that point on the expert system should be in full control of the assessment, and the engineer will be informed about the outcome at the end of the process. The process itself will consist of symbolic and numeric evaluations that will be combined for the final conclusions. Variations of that approach have been adopted in the two expert systems that are discussed next.

3.1 Safety assessment of concrete dams

The behaviour of concrete gravity dams can be represented by models for all factors involved in dam engineering. Such models are used for describing hydraulic conditions, foundation behaviour, structural response, and the interaction between these parameters. In all cases, that knowledge is procedural and quantitative. However, the inspection of existing dams is based on employing qualitative knowledge, rather than precise physical–mathematical models. The inspector uses his understanding of the observed problems instead of applying equations for evaluating it, and he may translate his interpretation into a form of input for a later quantitative analysis [7]. However, a qualitative evaluation could be subjective (i.e. two inspectors may provide different interpretations of the same observation), and one needs to develop a procedure for ensuring consistent results. This source of difficulties can be overcome by employing fuzzy set theories [4,24,29]. Such an approach, however, was not employed for this expert system; instead, this development concentrated on identifying and capturing the engineering reasoning by means of 'traditional' symbolic procedures for representing the data. To ensure consistency, the approach was tested and calibrated until the results were consistent with the real world. Another source of conflict may arise between the results from the complete assessment and the observed facts in the field. It is known that a dam can be in operation for many years before a weakening of the foundation is noticed [16] and the inspector will have to combine results from an accurate analysis with rational interpretation of present conditions for ensuring an acceptable, yet safe, conclusion.

In order to perform a risk analysis for a given dam, one needs to compare the dam's condition with an acceptable set of risk criteria [9]. The general approach for evaluating an observed condition has been illustrated in Fig. 3.1, and it consists of several options for obtaining an answer. For the present problem, the risk was chosen as the likelihood of uncontrolled water release caused by structural failure due to a combination of loading, support, and pre-existing conditions. The evaluation of these conditions is performed, initially, in a qualitative manner (i.e. by the symbolic part of the program) to obtain an estimate of the dam's stability. Then, if required by the user (and if the data for performing the calculations are available), the following four safety coefficients are calculated: ratio of resisting moment to overturning moment with respect to the toe of the dam; ratio of the compression at the toe to the allowable bearing pressure; ratio of downward pressure less the uplift pressure at the heel, to a threshold value of $100 \,\text{lbf in}^{-2}$ (4.8 kPa); the ratio of resisting to sliding forces at the dam foundation interface. These factors are then compiled to present the likelihood of dam stability problems which may cause an uncontrolled water release, as illustrated in Fig. 3.2.

The computation of the corresponding safety coefficients with respect to overturning and sliding instability is performed according to well-accepted principles [13,26]. The expert system performs the safety coefficient computation according to the engineering model and parameters provided by the user. It was assumed that both the model and the provided data are accurate, and the expert system treats the information with absolute certainty.

In that expert system [9–11], the combination of symbolic and numeric assessments is performed as follows: the assessment is only qualitative (i.e. symbolic), but a numeric evaluation of safety coefficients is performed if the user requires it. A qualitative evaluation cannot be performed without a field inspection of the dam; however, the user can perform a safety evaluation based on the known dimensions of the dam and the interpretation of safety coefficients. When both a symbolic and a numeric evaluation are performed, the results from the numeric analysis will govern in determining the potential problem of dam stability. Nevertheless, the expert

Ch. 3] ASSESSMENT IN A COMBINED SYMBOLIC–NUMERIC ENVIRONMENT 67

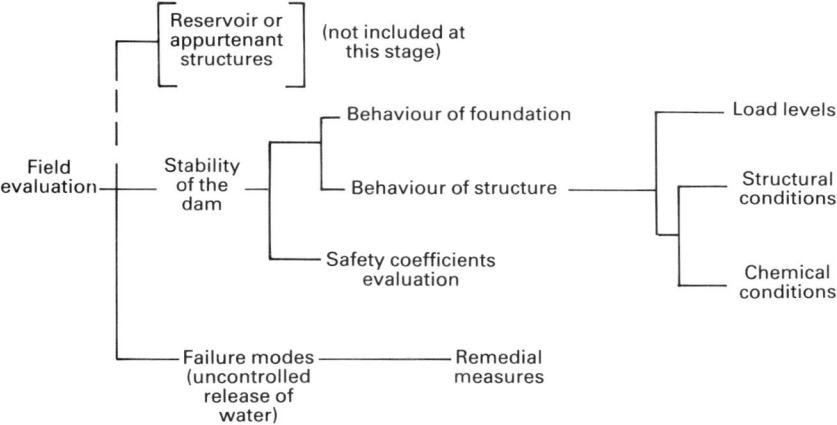

Fig. 3.2 — Combination of factors which affect the likelihood of dam stability problems.

system takes into account possible conflicts between results obtained by the numeric and symbolic approaches, and the level of concern based on the symbolic evaluation is raised or lowered depending on the direction indicated by the numeric assessment. The reverse influence (i.e. of the symbolic results on the numeric outcome) is assumed to be linear, based on the variation between the best and worst conditions for each case. When the results of the combined evaluation indicate a very high possibility for stability problems, the user will be asked by the expert system to check additional items in the dam. This will be required regardless of the results obtained initially by the numeric assessment.

The architecture for this expert system is illustrated in Fig. 3.3, while the structured knowledge base is illustrated in Fig. 3.4. The order by which the expert system shell starts and pursues the backward chaining process was used to define the order of the search. The root frame DAMEXPERT in Fig. 3.4 contains goal definitions according to first a parameter DAM-PROBLEM and second a parameter DAM-REPAIR. This ensures that the expert system will first try to identify the problem, and only then to propose remedial measures. Also, for this problem, shallow knowledge was derived from the literature and personal experience, and then it was organized according to deep knowledge from the literature and other engineering background on the design and behaviour of hydraulic structures. Finally, the capabilities of this expert system are illustrated by the addressing the following issues:

(1) provide an initial, qualitative evaluation of the support, loading, and pre-existing conditions on the dam safety;
(2) determine the values of four safety coefficients, as discussed above;

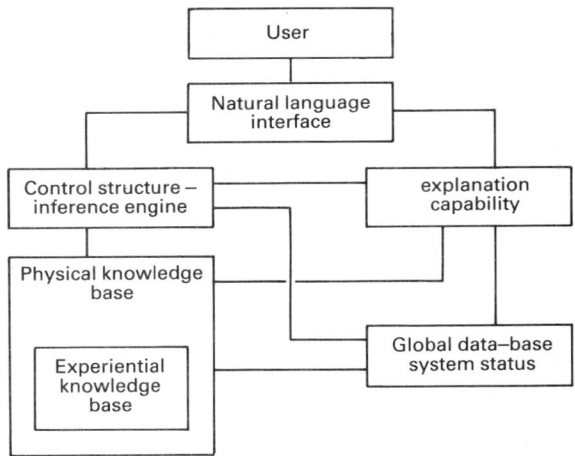

Fig. 3.3 — The architecture of the expert system for assessing dam stability.

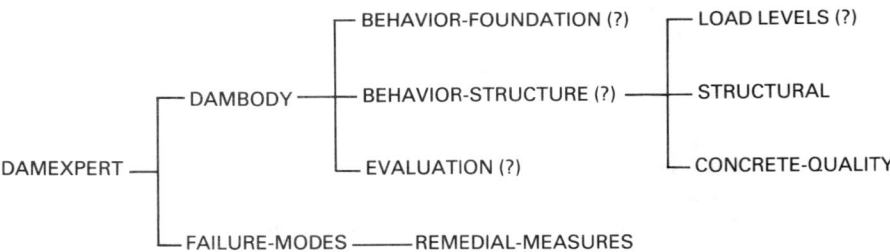

Fig. 3.4 — The structured knowledge base of the expert system for assessing dam stability.

(3) obtain a final evaluation of the possibility for a stability problem, by combining the results from the symbolic and numeric assessments;
(4) provide a confidence factor with respect to an uncontrolled release of water;
(5) list possible remedial measures for the observed problems;

(6) recommend additional inspections for supplemental data, if required;
(7) the option to review the data provided by the user, and the corresponding decisions reached by the expert system.

3.2 Evaluation of structures subjected to explosions

A general discussion of the problems related to this difficult area of structural engineering was presented previously [19], and here only a brief outline will be provided. The first issue to be addressed is the loading environments under which these structural systems are expected to function. In general, there are two such environments, the first associated with a nuclear detonation, as defined in the literature [2]. Such an environment would depend on the magnitude of the detonation, the relative position of the detonation centre with respect to the ground free surface and its distance from the structure, the position of the structure with respect to the ground free surface, geological conditions of the site, and several other factors. These parameters can be incorporated into relationships that were derived for the prediction of pressure–time histories and ground motions to which the structure would be exposed. The second environment is that associated with the domain of conventional detonation effects that could be generated by chemical explosives. Here the issue of environment definition seems to be somewhat more difficult to formulate [3,15,25]. In general, it is quite clear that one can measure detonation effects relatively far from the centre of an explosion, but as one approaches the centre the instrumentation may no longer survive to provide the required data. In the case of a nuclear detonation, that may not be a major problem since the structures are expected to function only if located outside of the induced crater; however, because of the large pressure gradients that are associated with conventional detonations the structures may have to be located considerably closer to the detonation centre in order to be extensively damaged. Several other significant differences between nuclear and conventional detonations which tend to vary the structural response were discussed by Krauthammer [18] and are briefly outlined next.

When an explosive device is detonated in the air, an airblast is applied to the soil free surface, and a corresponding shock wave is induced in the soil directly beneath it. That ground shock will propagate in the soil until it meets with the structure, and as a result of the soil–structure interaction the loading environment on the structure will be produced. The structural response is controlled by structural mechanisms which respond to the induced loads. Such mechanisms may be represented by direct shear which could cause failure at specific locations at a very early time, by flexure where a slab may exhibit membrane effects initially in compression and later possibly also in tension, by a compression mode in the event that the structure is an arch which is loaded in the vertical direction, or by combinations thereof. Also, it is quite clear that nuclear detonations, or very large chemical explosions, tend to load the entire structural system with relatively long pressure–time histories, while small conventional explosions may only affect local regions in the structure for considerably shorter durations.

Naturally, such differences in the loading functions will activate different modes of response in the structure, and they have to be explicitly addressed in the development of an expert system.

The present expert system is being developed to serve the following six functions:

(1) to design structural elements for protective construction;
(2) to assess structural damage, and to estimate future resistance capabilities;
(3) to identify the failure mode(s), and sequence of events;
(4) to advise on an appropriate method of numerical analysis, and to analyse structural elements under the effects of explosions;
(5) to recommend on how the structure can be stored, if possible;
(6) the option to review the user's input and the logic flow and decisions reached by the expert system.

The general sequence of damage assessment is the same as described by Krauthammer [19] and illustrated in Fig. 3.5, from which it can be understood that the expert system will assess each structure according to its geometry, burial conditions, and observed damage. For each type of structure, the system will perform an assessment starting with a decision on the nature of the loading condition (i.e. in order to determine whether the load was induced by a conventional or a nuclear detonation). The other issues to be considered are related to soil–structure interaction, the nature of the structural response, classification of damage, and an assessment of the reserve capacity. The output includes comments on each of the criteria that were employed for reaching decisions, and the user can ask to have these comments displayed.

The general inference tree for all the cases is illustrated in Fig. 3.6, which illustrate the logical flow that was implemented in the system. It can be noticed from Fig. 3.6 that the observations made on site are combined with any possible prior information on the particular structure that may exist in the database for gradually evaluating the problem under consideration. The expert system calls other programs, located on different machines, for performing parallel numerical evaluations of specific problems. The input for each program is prepared automatically from the data stored in the 'memory bank', also located on another computer, and the output is transferred back to the expert system as a batch file. The expert system, which continues to process other information, is notified that the numeric data has arrived, and will access it to conclude the operation on that problem. Thus the expert system continues the assessment procedure until a final decision is reached. From the decisions on the structural response one can reach a conclusion on the local or global nature of the failure. Also, site observations and the information on the structural response are combined for determining the nature of possible dynamic soil–structure interaction and loading conditions. Once all that information has been accumulated, the system will determine the damage condition in the structure and its inter-

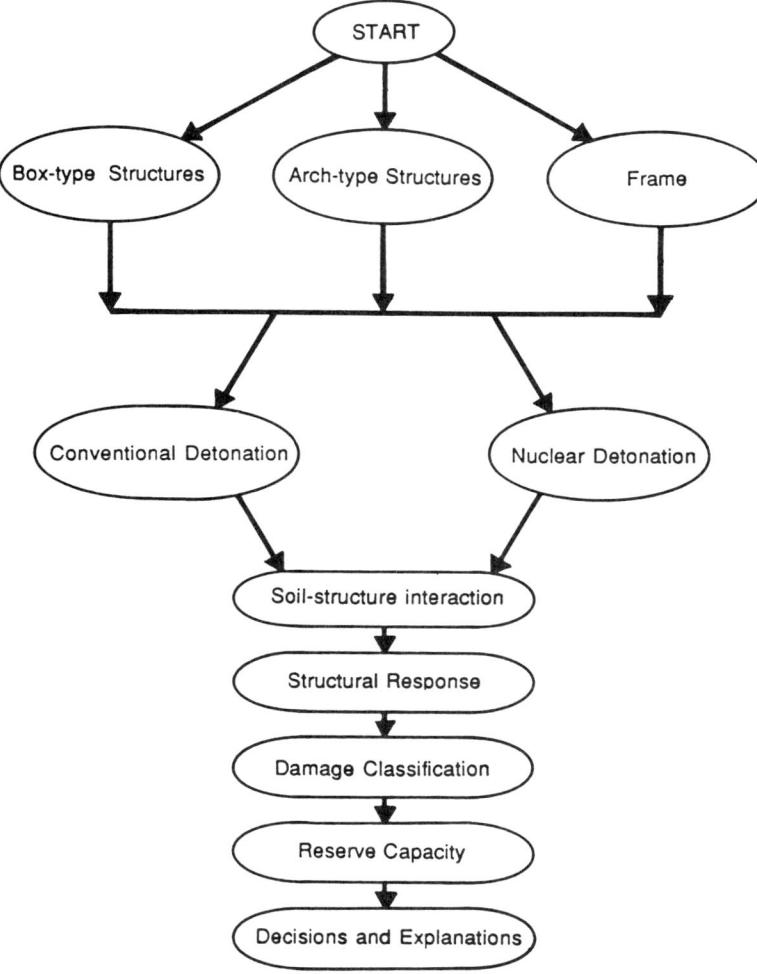

Fig. 3.5 — General sequence of detonation damage assessment.

face. Next the system can assess the reserve capacity of the facility, predict the future usefulness, and provide recommendations on possible corrective measures.

Similarly to the expert system developed by Wong et al. [28], the present expert system can be employed also for the design of future facilities and their assessment under possible loading conditions, and on the basis of the results to return to the design module for required structural changes. It is envisioned that such systems may be used in the future for the whole range of structural engineering activities from facility planning through the possible phase of restoration.

72 ASSESSMENT OF STRUCTURAL BEHAVIOUR [Pt. I

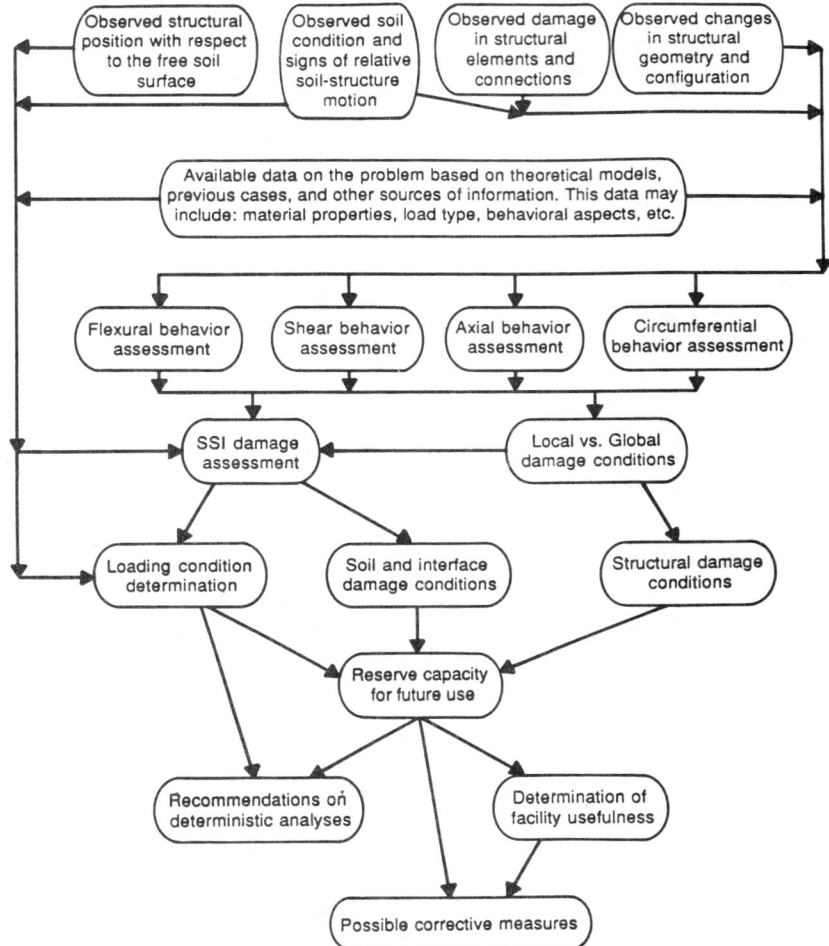

Fig. 3.6 — General inference tree for detonation damage assessment.

4. CONCLUSIONS AND RECOMMENDATIONS

This chapter was aimed at exploring the feasibility and effectiveness of combining symbolic and numeric approaches for the general task of structural assessment. It has been shown that such combinations are very desirable for they harness the power of two different approaches for the task at hand. Many problems cannot be efficiently treated separately by these approaches; however, when combined into one process they complement each other. Several expert systems based on this general concept have been developed, and others are under development. The lessons learned from each case are very important for determining the future direction in this field, and the present discussion of two such expert systems was aimed at that goal.

There is no doubt that coupled symbol–numeric programs are an important step in the right direction. Nevertheless, the reservations expressed by many about the danger of having unexperienced individuals operate expert systems for performing complicated evaluations hold true also for these advanced codes. There is no such thing as a 'perfect expert system'. Most existing expert systems are primarily experimental tools, and the decisions obtained by them have to be studied carefully by experts. The most useful purpose that expert systems can have is to serve as 'consultants' to experienced engineers for ensuring that all facts are considered, and that the final decision has been carefully examined. As we gain more knowledge on how to develop better expert systems, and learn to capture knowledge into symbolic and numeric modules, the role of expert systems in structural engineering will expand.

REFERENCES

[1] Adeli, H., *Expert Systems in Construction and Structural Engineering*, Chapman and Hall, London, 1988.
[2] ASCE, *Design of Structures to Resist Nuclear Weapon Effects*, American Society of Civil Engineers, Manual 42, 1985.
[3] Baker, W. E., Cox, P. A., Westin, P. S., Kulstz, J. J. and Strehlow, R. A., *Explosion Hazards and Evaluation*, Elsevier, Amsterdam, 1983.
[4] Blockley, D., Uncertain inference in knowledge based systems, *ASCE J. Eng. Mech.*, **113** (4), April 1987.
[5] Buchanan, B. G. and Shortliffe, E. H., *Rule-Based Expert Systems*, Addison-Wesley, Reading, MA, 1983.
[6] Carson, J. M., Ross, T.J., Hyndman, D., and Wong, F. S., Pattern recognition techniques for distinguishing structural failure modes from high pressure loading records, *Proc. 4th Conference on Computing in Civil Engineering, October 1986,* American Society of Civil Engineers, pp. 699–713.
[7] Clancey, W. J., Heuristic classification, *J. Artif. Intell.*, **27** (2), 1985.
[8] Degroff, L., Conventional languages and expert systems, *AI Expert,* April 1987.
[9] Frank, B. M. and Krauthammer, T., Preliminary safety and risk assessment for existing hydraulic structures — an expert system approach, *Structural Engineering Rep. ST-87-05,* Department of Civil and Mineral Engineering, University of Minnesota, August 1987.
[10] Franck, B. M. and Krauthammer, T., Development of an expert system for preliminary risk assessment of existing concrete dams, *Eng. Comput.*, **3**, 137–148, 1988.
[11] Franck, B. M. and Krauthammer, T., An expert system for field inspection of concrete dams: part 1, engineering knowledge, *Eng. Comput.,* to be published.
[12] Franck, B. M. and Krauthammer, T., An expert system for field inspection of concrete dams: part 2, artificial intelligence issues, *Eng. Comput.,* to be published.

[13] Golze, A. R., *Handbook of Dam Engineering,* Van Nostrand-Reinhold, New York, 1977.
[14] Hayes-Roth, F., Waterman, D. A. and Lenat, D. B., *Building Expert Systems,* Addison-Wesley, Reading, MA, 1983.
[15] Henrych, J., *The Dynamics of Explosions and Its Use,* Elsevier, Amsterdam, 1979.
[16] Jansen, R. B., Carlson, R. W. and Wilson, E. L., Diagnostic and treatment of dams, *Trans. Madrid Congress,* International Commission of Large Dams, 1973.
[17] Krauthammer, T. and Kohler, S., RC structures under severe loads — an expert system approach, *Proc. ASCE Symposium on Expert Systems in Civil Engineering,* American Society of Civil Engineers, 1986, pp. 96–108.
[18] Krauthammer, T., A numerical gauge for structural assessment, *Shock Vibr. Bull.* **56**, 179–193, 1986.
[19] Krauthammer, T., Damage: an expert system for structural assessment and failure identification following blast and shock loading events, *Eng. Comput.,* **3**, 69–86, 1987.
[20] Krauthammer, T., Analysis of reinforced concrete structures under the effects of localized detonations, *Shock Vibr. Bull.,* **57**, 9–18, 1987.
[21] Krauthammer, T., Shahriar, S. and Shanaa, H. M., Analysis of reinforced concrete beams subjected to severe concentrated loads, *ACI Struct. J.,* **84** (6), 473–480, November–December 1987.
[22] Rich, E., *Artificial Intelligence,* McGraw-Hill, New York, 1983.
[23] Ross, T. J., Wong, F. S., Savage, S. J. and Sorensen, H. C., DAPS: an expert system for damage assessment of protective structures, *Proc. Symp. on Expert Systems in Civil Engineering,* American Society of Civil Engineers, 1986, pp. 109–120.
[24] Ross, T. J., Approximate reasoning in structural damage assessment. In Adeli, H. (ed.), *Expert Systems in Construction and Structural Engineering,* Chapman and Hall, London, 1988, pp. 161–192.
[25] United States Department of the Army, Fundamentals of protective design for conventional weapons, *TM 5-855-1,* November 1986.
[26] United States Department of the Interior, Bureau of Reclamation, *Design of Small Dams,* Water Resources Publication, 2nd edn, revised reprint, 1977.
[27] Waterman, D. A., *A Guide to Expert Systems,* Addison-Wesley, Reading, MA, 1986.
[28] Wong, F. S., Dong, W. M. and Blanks, M., An integrated PC-based computer system for protective structures application, *Final Rep. TR-88-91,* Air Force Weapons Laboratory, 1988.
[29] Yao, J. T. P., *Safety and Reliability of Existing Structures,* Pitman, London 1985.

4

A blackboard consultation system for constitutive modelling in solid mechanics

J. R. Ambroziak and **M. Kleiber**
Polish Academy of Sciences

1. CONSTITUTIVE MODELLING AND CONMOD

1.1 Constitutive modelling

The work reported here has been done in the context of designing and implementing a consultation system in the field of continuum mechanics particularly in *constitutive modelling* [1].

The predictive value of any inelastic analysis depends on the following factors:

— constitutive law
  — required behaviour under selected load histories,
  — identification of material functions,
  — sensitivity to parameter variation;
— the way the boundary value problem is posed;
— solution algorithm.

In every numerical analysis of the behaviour under load of any mechanical system, a mathematical model of the material has to be employed. The model, referred to as the constitutive model of the material, covers its expected major physical properties while disregarding those of minor relevance.

Constitutive models form a big large taxonomy. Assigning them to solid mechanics problems is a knowledge-intensive activity with a markedly heuristic quality because of the empiric nature of the models and sometimes their provisional character for want of better models yet to be proposed. Each model is characterized by a set of equations. The better the match, the closer the correspondence between the results of numeric processing and physical reality. Usually mechanical engineers are not able to find optimal models and use overly general models, compensating model inadequately by parameter tuning.

The term 'constitutive modelling' has in solid mechanics the conventional meaning of the task of constructing the best possible model representing the thermomechanical behaviour of a given material. Here, a slightly different meaning of this term is adopted and constitutive modelling is understood as a selection from the existing variety of constitutive models of an 'optimal' one under specified constraints. The constraints are grouped in three categories, each of them representing:

(a) data concerning the overall engineering situation to be studied;
(b) data concerning those features of the material behaviour which are presumed to be of significance;
(c) data concerning existing software and hardware resources (finite element codes available, computer facilities, etc.) including indications as to the acceptable cost of the analysis.

According to the above scheme, in the course of an interactive session with CONMOD the user is encouraged to answer questions related to such topics as

(a) — general field of application of the model,
— desired level of sophistication of the analysis,
— philosophy of analysis,
— problem type,
— load type,
— dimensionality of the problem,
— interacting fields;

(b) — type of material,
— possibility of plastic effects,
— possibility of viscous effects,
— possibility of fracture effects,
— material symmetries.

The characteristics related to (c) have not yet been taken into account of in the current version of CONMOD.

Each of the above options splits into a number of more specific choices and so on, thus forming a problem description hierarchy. For instance, 'desired level of sophistication of the analysis' is currently followed by just two options:

— simplified analysis (traditional design),
— sophisticated assessment of the process,

whereas 'philosophy of analysis' implies the selection of one or more possibilities from the following list:

— complete analysis (step by step if time dependent),
— limit analysis,
— analysis at critical points,
— post-critical behaviour,
— fracture analysis,
— free vibration analysis (via the eigenvalue approach),

— forced vibration analysis,
— dynamic stability analysis,
— coupled fields analysis,
— unilateral (contact) problems,
— optimization,
— sensitivity assessment.

Similarly, 'problem type' splits into:

— purely mechanical,
— coupled,
— purely thermal,
— other,

with 'coupled' followed by:

— thermomechanical,
— fluid–structure interaction.

1.2 Aims of the project

The system called CONMOD is designed to find, in an interactive session, an appropriate model of the material with accompanying equations to match the problem description given by the user. Attention is paid to gather all relevant data and to enforce consistency among them, i.e. to eliminate data that, collectively, do not make physical sense.

CONMOD will be used for consultation so the quality of user–system dialogues is crucial. Since in addition to reasoning, handling graphics I/O and possibly numeric computations call for a sufficiently general and extendable programming environment, we decided to employ the blackboard architecture [2].

The primary means of deductive inference in CONMOD provides the resolution rule [3]. In order to support different directions of the consultation, we prefer to regard our rules as constraints between facts rather than implications. Specifically, it is also practical to derive material features from an assumed model. In contrast to strictly forward or backward chaining, the resolution rule makes it possible to use clauses in either direction.

This generality normally leads to a combinatorial explosion of spurious consequents (resolvents). Setting the resolution rule in the blackboard framework makes it possible to exploit the generality of the rule while avoiding the combinatorial explosion. This is achieved by implementing appropriate reasoning control strategies at the blackboard level.

In the long term we plan CONMOD to be capable of conducting an extended consultation: answering questions about constitutive modelling in general, comparing models, finding materials to match a given model, etc.

2. THE BB FRAMEWORK

The general blackboard programming framework, called BB, based on the concept proposed by Hayes–Roth [4] was written in the C language, initially for IBM PC computers, and has already been transported to a SUN 3 workstation. Apart from CONMOD, the system is being used in a medical monitoring application, and is planned to be developed as a sophisticated expert system shell.

The program can be considered to be an interpreter that reads in several files containing symbolic descriptions of various declarative and procedural modules, and then runs the latter according to the architecture's control principles. The design of the BB system includes a set of syntactic conventions forming a specialized language (BBL), in which primitive and complex data structures, as well as expressions to test and transform the data, can be written.

Because of the increasing popularity of the architecture its essentials are widely known, and it is used in many variants [5]. In the following description we will present only those features of BB that are referred to in subsequent sections.

2.1 Basic building blocks

The fundamental symbolic data structure of BB is a structured object — like the LISP atom — with a name and a property list. All data on the blackboard, where problem-solving activity take place, are implemented as structured objects.

Essentially, there are two families of objects: system objects and user objects. System objects are used by the underlying implementation and have rigid forms; they are closely connected with the control structure of the blackboard style of information processing. On the other hand, user objects can have arbitrary attributes; it is up to the user's creativity to design a taxonomy of objects to serve useful roles in the processing.

Also, the knowledge sources (KSs), which can be viewed as generalized if–then rules or complex, event-driven procedures responsible for information processing, are likewise implemented. As objects, they consist of about 12 property–value pairs, containing:

— descriptive data,
— two procedural tests of the KS applicability,
— a piece of code to estimate data utility,
— scheduling data,
— the main procedural body.

All procedures and tests are assembled from primitive expressions which are implemented directly in the C code. For the time being, new primitives cannot be defined on the symbolic level; the required procedures have to be written in the implementation language. For efficiency, some complex operations that could in principle be expressed by elementary expressions are implemented as primitives. This applies specifically to the operations responsible for resolution inference. In particular, a single primitive called resolve computes the resolvent given two appropriate clauses.

2.2 Functioning of BB

The system operates in cycles. Progress is achieved through blackboard changes, namely object creations and/or modifications. Modified objects are put on an event list. As events, they trigger KSs, i.e. the trigger test of each KS is executed for each element of the list.

To speed up execution, an obvious modification of the straightforward algorithm has been introduced. Each KS explicitly declares through one of its attributes the type of objects which it can process. After having been read, all KSs are organized in several lists according to their type. Modified objects (events) are in each cycle associated with corresponding lists. This means that modifications of objects of any type for which there are no KSs are simply ignored. Then, trigger checking is done separately for each type.

When the test succeeds, it roughly means that the KS expression applied to the event can potentially contribute to the problem solution. At this moment a special object called the knowledge source activation record (KSAR) is created and added to the triggered list. The KSAR stores, among other things, information about the triggering event, the triggered KS, and variable bindings made during trigger testing.

Once all new KSARs have been recorded, the elements of the triggered list are scanned once more to check whether the second of the KS tests — the precondition — succeeds for the KSAR data as well. While the trigger test concerns event structure, the precondition is sensitive to a global blackboard state, e.g. to the problem-solving phase, or to the objects's changing characteristics, e.g. whether it is currently believed or not. Thanks to the precondition, the actual execution of any suggested activity can be delayed until some global conditions hold, or until the object is in a 'proper' state. This delayed execution differentiates blackboard architecture from that of a forward rule system and brings much of its power.

When for a given KSAR the precondition succeeds, the KSAR is moved from the triggered list to the invokable list. At the same time, another of the KS's expressions, the immediate code (named after a similar construct in Hearsay III [6]), is run to produce a numeric result interpreted as the KS's own estimate of the importance of the data to be processed. For instance, in deductive reasoning using the resolution rule contradictions manifest as empty clauses (this is in contrast to the refutation theorem proving, where the empty clause marks completion of a proof), i.e. clauses containing no literals. As the early recognition of the contradictive data is very important, a sufficiently high immediate code rating will be assigned to any pair of clauses about to resolve to an empty clause. In this case, the immediate code is a function sensitive to the number of literals in candidate clauses, computing diminishing numbers with growing number of literals.

Finally, elements of the invokable list are rated against current global scheduling criteria, where the immediate code value is an important factor, and the highest rated KSAR is chosen for execution. In the process, the procedural body of the KS is evaluated in the environment of variable bindings made during testing — bringing about the next generation of events for the next cycle. The system terminates execution in reponse to the halt primitive, or whenever the invocable list becomes empty.

Trigger tests should correspond to fixed object characteristics; they are responsible for recognizing objects that the KS can and should process. We can threat knowledge sources as one-parameter procedures created or modified objects are substituted for the parameter before executing the KS's procedural atibutes. The substituted object need not be tested for its' type — the test is implicit in the control structure. Normally, other of the object's attributes are tested: whether they are defined and/or whether they have specific values. For example, the KS responsible for processing atomic facts tests for clauses containing exactly one literal. Such characteristics define the object and are fixed.

On the other hand, as mentioned earlier, preconditions should concern only charging state and/or object characteristics. For example, all KSs responsible for deductive inference insist in their preconditions that the fact base be contradiction free. When a contradiction occurs, all pending inferences are suspended; they can depend on contradictory data themselves. After the reason maintenance system gets rid of the contradiction and the clauses taking part in the suspended inferences remain believed in, the conclusions can safely be made and asserted.

Another example of the precondition's application concerns concurrent programming. In CONMOD there are objects called demons which check facts sent to them. Although there can be many such facts, demons can process them only one by one. To organize the processing, demons are equipped with special attributes to keep their local state; in particular, the currently considered single fact becomes associated with the demon. When the demon is 'busy', all activities trying to make it consider new facts 'wait': the corresponding preconditions states that, to accept a new fact, a demon has to be 'free'. After the demon becomes free, one — the highest rated — activity succeeds in giving the demon its fact for consideration, so the demon becomes busy again and the other KSARs have to wait further.

Writing the application on the knowledge level resembles classic programming very much. Firstly, KSs' codes are normal, sequential procedures with assignments, conditional branches, iterative loops, etc. Even if — as in our case — KSs can also contain if–then rules, the rules are used in a strictly designed combination of iteration and testing. Secondly, although the power of the blackboard architecture lies in its opportunistic non-determinism and inherent parallelism, all interactions between the KSs should likewise be consciously, carefully and explicitly planned. In particular, some KSs produce objects that are best understood as messages, i.e. they do not contribute to the problem solution directly, but are specifically assembled to trigger some KS to stimulate further processing. Similarly, the designers know for certain which object characteristics are defining, and which are additional, volatile data, so there should be no doubt as to what to test in the trigger and what in the precondition.

3. THE KNOWLEDGE BASE

3.1 The knowledge sources

The BB–CONMOD knowledge base consists of two main parts. The first of these contains KS descriptions as well as two procedures, called Prolog and Epilog, to be unconditionally executed respectively before and after KS interpretation.

The Prolog procedure initializes the BB environment and evokes the starting event. In the constitutive modelling consultation, Prolog assembles an object of the **Goal** class, pointing to the root of the problem description.

The set of KSs can be viewed as a modular and explicit inference engine to be incrementally developed. The main processes being supported, new KSs can be added to complement the basic activities and to increase the global efficiency of the system. The CONMOD interpreter consisting of an algorithmic interpreter and a set of knowledge sources is schematically illustrated in Fig. 4.1.

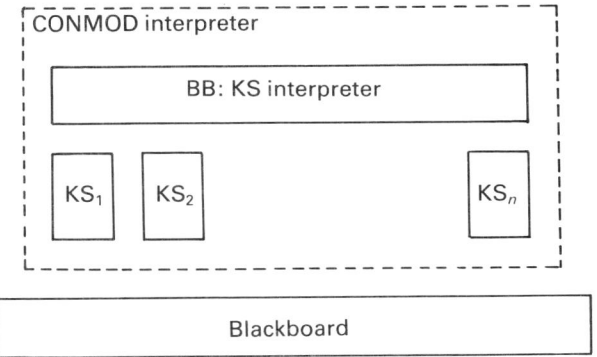

Fig. 4.1 — CONMOD interpreter.

3.2 The UNITS

The second part of knowledge base, called UNITS, contains definitions of problem description constituents, predicates and constants, as well as constraints — logical sentences in object form. In UNITS, English translations of domain-dependent terms, possible values of the predicates, etc., can be found. The definitions make use of only domain-independent property names.

It should be pointed out that, in contrast to the UNITS, the KSs are completely domain independent, even if they refer to the properties of the units.

4. REPRESENTING LOGICAL SENTENCES

To embed the resolution rule in the above framework we have designed a mapping of resolution concepts into blackboard notions.

Most importantly, we propose a notation for representing clauses as a special kind of structured objects. It has been developed in response to dissatisfaction with our previous means of representing facts as structured objects.

4.1 Structured objects

In the simplest case we can store information about any concrete or abstract object as a set of attribute–value pairs. We can add more and more characteristics state elementary facts and the structured object is implicitly interpreted as a conjunction of the facts.

The notation presents some difficulties. It is not clear how to provide independent justifications for atomic facts when they are added by different rules. Similarly, it is not evident how to represent negation and contradiction information.

These problems can be solved by representing facts explicitly:

```
Fact_n
    type:              fact
    object:            ⟨object of the fact⟩
    isa:               ⟨class of the object⟩
    ⟨attribute⟩:       ⟨value⟩
    ...
    status:            ⟨believed or not⟩
    justification:     ⟨justification⟩
```

Each object is interpreted as a conjunction of elementary facts about its object, but we can have several facts concerned with the a single entity. The reformulation provides many useful possibilities for knowledge representation, e.g. we can represent beliefs of different agents by objects referring to sets of fact-objects.

However, the form suffers from at least two deficiencies:

(1) it involves the proliferation of domain-dependent slots, and the reasoning procedure must be informed about their names and meaning (usage);
(2) it offers no efficient indexing scheme.

4.2 Prefix predicate calculus notation

Looking for a computationally efficient notation with adequate expressive power, we also considered the benefits of the traditional list notation of predicate calculus formulas.

List patterns suit the needs of knowledge processing very well [7]. They are capable of expressing complex information using recursively defined term syntax. In fact, list patterns correspond to tree structures while objects are 'one dimensional'. Information in pattern form can be efficiently indexed in a discrimination tree [8,9], and the reasoning mechanism benefits from natural application of the unification procedure. Logical sentences can be annotated with arbitrary information (uncertainty, justifications, etc.).

However, some problems emerge with respect to complex formulas. How should they be indexed? Should subformulas be used as keys? What about unification pattern matching? Such questions arise because of the loss of definite structure in compound formulas: the number and order of subformulas are variable.

4.3 Putting sentences in object form

These considerations have led us finally to what we call a pattern–object form (POF), a synthesis of structured object and list pattern notations, preserving their merits and disposing with their above-mentioned drawbacks.

The POF idea consists in separating — for each formula — its constituent atomic formulas from its logical structure, and storing these data in a structured object. For instance the formula:

$$(\text{PhilAnalysis LinearAnalysis}) \wedge (\text{ProblemType Thermal}) \Rightarrow (\text{ModelClass LinearHeatConduction})$$

(where PhilAnalysis denotes philosophy of analysis) can be written as follows:

```
Formula_10
   type:     complex_formula
   a:        (PhilAnalysis LinearAnalysis)
   b:        (ProblemType Thermal)
   c:        (modelClass LinearHeatConduction)
   schema:   (if (and a b) c)
```

Properties a, b and c are called **roles**, and are used for cross-reference within the object. The **schema** property describes, using the roles and the usual logical connectives, the structure of the formula.

Formulas in POF are free from the inconveniences mentioned above. All predicate calculus formulas can be represented in a uniform way. Each object corresponds to a single formula so there is no need for partial justifications. Domain-dependent notions form the vocabulary of predicate calculus constants and do not influence object property names. It should also be noted that the POF is convenient for knowledge acquisition — formulas are easy to write down.

The inference engine uses unification without the difficulties caused by variable number and order of atomic formulas. In fact, the POF roles are used to represent the set of atomic formulas, and sequence characteristics that were necessarily important for the tree pattern structure are in POF abstracted from (all the same, such information can be restored from a schema). Unification is used only to match atomic formulas. The BB system provides primitives for matching and building atomic formula patterns.

4.4 Representing clauses

The next step taken in the BB system is the transformation of formulas into the conjunctive normal form (CNF). This is done automatically on reading the formulas from the UNITS base. When a given formula is not expressible as a disjunction of literals but as a conjunction of clauses, a new object, to be justified by the original formula, is created for each conjunct. The collection of all believed clause–objects is interpreted as a conjuction of clauses represented by the objects.

When the logical schema of a sentence is expressed in the CNF, the resulting clauses can be represented in a still simpler form. It is sufficient to mention the set of atomic constituents of the clause and to differentiate negated literals from non-negated literals.

In order to discriminate between negated and non-negated constituent literals new properties **pos** and **neg** are introduced. Their values are set to lists of roles corresponding to respective literals. As

$$(a \wedge b \Rightarrow c) \equiv (\sim a \vee \sim b \vee c)$$

the example object becomes:

```
Formula_10
   type:     complex_formula
   a:        (PhilAnalysis LinearAnalysis)
   b:        (ProblemType Thermal)
   c:        (ModelClass LinearHearConduction)
   schema:   (if (and a b) c)
   neg:      (a b)
   pos:      (c)
```

4.5 Indexing of sentences

Formulas in object notation are multiply indexed in the discrimination tree using patterns of atomic formulas as keys. Negated and non-negated literals are stored separately, as they are mainly retrieved when looking for complementary literals to a given literal.

For instance, the following calls (here in a slightly stylized form)

```
        index(sign:neg,data:(Formula_10 . a),
                       key:(PhilAnalysis LinearAnalysis))
        index(sign:neg,data:(Formula_10 . b),
                       kay:(ProblemType Thermal))
        index(sign:pos,data:(Formula_10 . c),
                       key:(ModelClass LinearHeatConduction))
```

are used to index the example Formula_10 object.

Indexed formulas are retrieved by **fetch** (*sign*, *pattern*) which returns list of dotted pairs of the form ((*object. property*) ...). For each element of the list, get(car(*el*), cdr(*el*)) returns the actual pattern to be unified with the second argument of **fetch**.

Indexing sentences has two applications:

(1) it helps to find clauses to resolve with a given clause through the complementary literals — easily accessible in the discrimination tree;
(2) it is used in the **uniquification** procedure, which checks for each clause about to be asserted whether it has already been known, in which case assertion amounts to adding a new justification to the found object.

5. RESOLUTION INFERENCE

5.1 Knowledge source implementation of the resolution

When a new clause–object is created, it triggers — as an event — the **Find candidates** KS responsible for finding clauses to resolve with the given clause. All clauses are accessible via the patterns of their atomic formulas so the task is done without difficulty. However, to eliminate many redundant resolvents, only those resolution possibilities are noted that fit the ordered resolution strategy [3].

The possible inferences are not made immediately; the KS produces a set of objects called resolve pairs. Each of them points to two clauses and specifies roles of the literals to be resolved on.

New resolve pairs trigger other KSs. One of them — **Resolve** KS — using resolve pair's data actually builds the resolvent, computes its justification and the certainty factor, and finally asserts (an asserted resolvent generates a new event).

These two KSs are responsible for the main resolution 'loop'. They use specialized primitive actions designed to speed up execution.

5.2 Controlling the resolution process

The immediate code of the **Resolve** KS estimates the utility of the would-be resolvent. At present, it assigns the maximal value to an empty clause, and values decreasing with length (number of literals) to others. Unit resolution is implemented through the KS's trigger that succeeds only for resolve pairs pointing to at least one one-literal clause.

Two points are worth making here. Firstly, the combinatorial explosion of redundant resolvents is avoided when the resolution process is controlled by the blackboard mechanism with its philosophy of recognizing opportunities for action but delaying the actual execution (or even abandoning it altogether) until its rationality has been confirmed by the KS tests, and it has been assigned a sufficiently high rating. Moreover, only clauses created during the session can have consequences, because, as events, only they will be spotted; this can be termed event-driven resolution.

Secondly, the same event can trigger several KSs so that more than one activity concerning a created or modified object can be examined. We have exploited this opportunity to prune would-be resolvents when considered redundant. In addition to triggering the **Resolve** KS, each resolve pair triggers the **PruneResolvePairs** KS of higher static priority (a KS characteristic influencing KSAR rating). The KS checks resolve pairs in the context of all currently believed clauses and deletes redundant resolve pairs — at present, those that can be interpreted as implications (rules) with no chance of antecedent satisfaction.

For each deleted object, the BB mechanism takes care of deleting all KSARs concerned with the object. With resolve pairs, it means that some possible resolution inferences are cancelled by **PruneResolvePairs** and never actually made.

5.3 Associated activities

Newly created clauses trigger also other KSs. One of them checks whether a given clause is a unit clause and, if so, it displays to the user the clause's contents in English (using information from the UNITS).

Another KS complements resolution with elimination for clauses consisting of a single negated literal. The KS uses the UNITS data specifying possible values of the formula's predicate. When the predicate has two possible values and the input clause excludes one of them, the **Eliminate** KS asserts the other as the value of the predicate.

5.4 Contradictions

If an empty clause is 'frozen' in some resolve pair it will be created in the first place. The derivation of an empty clause is a manifestation of a contradiction, and hence of the inconsistency of the currently believed set of clauses. When a contradiction appears, the global blackboard flag, **Contradiction** is set. It is tested by the preconditions of several KSs and, the flag being set, all pending inferences are blocked.

The BB system uses a Doyle-type TMS [8–11]. As a user is likely to enter data that do not correspond to the solid mechanics point of view (e.g. by not knowing the exact scope of some definition), measures must be taken to recover from such failures. Every fact coming from the user is positively justified by the user's stating it, and negatively justified (with an out-justifier) by his possibly making a mistake.

When an empty clause has been derived, resolution stops and data dependency backtracking begins. A dedicated KS finds the set of assumptions leading to the contradiction, chooses one and makes it OUT by justifying its out-justifer (supporting the view that he did make a mistake). After the contradiction has become OUT, in the process of label propagation normal activity is resumed.

6. MODEL SELECTION

The BB–CONMOD system is still under development. The ideas we have adopted have an experimental status; whether they survive to subsequent versions of the tool–knowledge base, only time will show.

6.1 Knowledge source driven consultation

In the constitutive modelling consultation, the Prolog module assembles an object of the **Goal** class, pointing to the root of the problem description; see section 1.1. The emergence of the goal triggers the Decompose KS, which creates offspring subgoals. These being **Goal**s themselves are — in a recursive manner — similarly decomposed unless they happen to be primitive goals, i.e. single attributes as opposed to structured descriptions. If this is the case, they are used to generate questions concerning the problem at hand.

User answers, also in object form, are transformed to clauses that trigger resolution and activate watches as explained in section 6.3. Some of the resolution conclusions state the relevancy of further, i.e. beyond the scope of the tree in section 1.1, attributes of the problem (follow-up questions), e.g.

(relevantAttribute Kinematics MagnitudeOfStrains)

Such clauses trigger one of the KSs to produce new 'investigate' goals.

6.2 Representation of the constitutive models

The actual constitutive models are represented as structured objects arranged in the **general/specific** lattice. Any given object can have several submodels; more specific versions to be applied when the problem situation is known in finer detail or when more specific assumptions can be made. The model can also have more than one generalization so the graph representing the relationships between the models is not a tree.

For the time being, the models we have represented have only a couple of defining (i.e. not changing during the consultation, constituting the meaning of the object) attributes. For instance, the following object represents the linear thermoelastic material model:

```
LinearThermoElastic
   type:          ConstitutiveModel
   more_general:  (NonlinearThermoElastic ThermoElastoPlastic)
```

```
more_specific:  (IsotropicLinearThermoElastic LinearElastic)
verbal_descr:   "A basic model describing thermo-mechanical
                 coupling"
complexity:     3
```

The **complexity** is understood as a measure of computational effort needed to solve a typical problem involving this model. It is described with reference to an approximation scale defined by the **LinearElastic** model's having complexity of 1 and say, the **LargeStrainThermoElastoVisco Plastic** model's having a complexity of 10. The parameter serves to estimate the possible cost of the analysis using the model, as well as a cut-off criterion to eliminate selection of costly models when the user declares he wants a simplified model.

After the model selection issues are satisfactorily implemented, we plan to store data about the models in a hypertext system [12]. Hypertext is just what is needed in constitutive modelling, as the models have many meaningful relations between one another, and with each of them data of various forms can be associated — verbal and pictorial. The hypertext system should be a standalone program to be used independently of CONMOD for browsing through the realm of the models. At the same time, CONMOD should be interfaced to the hypertext, so that the consultation would end up in a 'guided tour' through the models or, to put it differently, the hypertext facility would constitute a rich form of the consulation results exposition.

6.3 Matching models

To connect the data-gathering part of the consultation with model selection — the goal of the consultation — two specialized types of structured objects have been introduced: **watches** and **demons**.

The main idea is that the facts gathered from the user or inferred in resolution or by elimination be spotted by watches — objects built around and indexed by fact patterns. After each act of fact recognition, a watch notifies objects associated with it called demons — objects capable of procedurally testing watch data, and performing arbitrary computations. In the model selection procedure, demon activity consists in assigning evidence characteristics to specific models. The flow of activations is depicted in Fig. 4.2.

Each piece of evidence consists of the demon name (for explanation and bookkeeping purposes) and a datum taken from the following four categories:

(1) a positive number: evidence for a model;
(2) a negative number: evidence against a model;
(3) the symbol **accept**: definite evidence for a model;
(4) the symbol **reject**: definite evidence against a model.

The numeric evidence data of a model are summed to produce the total evidence figure, which is compared with the threshold value. The result of

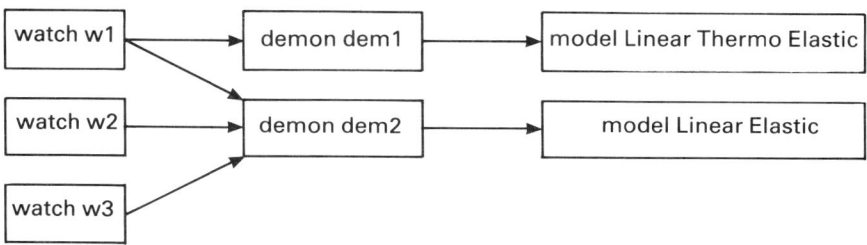

Fig. 4.2 — Model activation.

the comparison can be interpreted as definite evidence for or against a model, or only as a degree of confidence in the model.

In addition, whenever new evidence data are attached to a model the data are propagated in part to the submodels of the model as their inherited evidence. During the consultation direct and inherited evidence data accumulate in models. As the results of the consultation a list of models sorted by their evidence sums is displayed.

6.4 Gathering evidence

An example of a watch in its actual form from the UNITS knowledge base is

```
object {name:    w1
       type:     watch
       pattern:  (MaterialType ?mt)
       demons:   (dem1)
       }
```

The watch will be activated by the emergence of facts of the following form:

```
Formula_11
    type:   complex_formula
    a:      (MaterialType Metal)
    pos:    (a)
```

where in the unification process the variable ?mt will be bound to **Metal**.

The most important attribute of a watch is its **pattern**; every watch has a single pattern, which can contain variables. In the process of reading in the knowledge base, watches are stored in the discrimination tree according to their patterns. Watches are used only in connection with demons; each 'works for' at least one demon.

Similarly, demons can use more than one watch. An example of an actual demon from the CONMOD database is

```
object {name:      dem1
        type:      demon
        watches:   (w1)
        vars:      (?mt)
        test:      [member ?mt (GeoEng Biomech)]
        action:    [set LinearThermoElastic:evidence
                       [cons [list 'dem1
                                   [if [= ?mt 'GeoEng]
                                       [value −6]
                                       [value 'reject]]]
                             LinearThermoElastic:evidence]]
}
```

The object carries the information that the **LinearThermoElastic** model is not likely to occur in geotechnical and biomechanical problems; in the case of a geotechnical problem, it receives a negative score, but the model inapplicability is not as definite as in the other case.

test and **action** are the principal attributes of a demon; the names explain their roles. Both are expressions in the BBL language. The expressions can contain variables from the demon's watches' patterns. All such variables are listed in a separate attribute of a demon. The actual program for demon processing is rather complicated as it takes concurrency into account and has to deal with the combinatorics of sets of facts matching each of the demon's watches. The basic idea is as follows. When a demon is activated through a message from source of its watches, it checks whether each of them has matched any facts. (In fact a dedicated KS does the processing, but it seems easier to describe the process as if a demon were the active agent.) If the condition is confirmed, the facts are unified with the patterns of the watches in order to bind pattern variables with fact terms. When, in the process, all the demon's variables become bound, the **test** is executed, and if it evaluates to a non-**Nil** value the **action** evaluation follows. It can contain arbitrary code; however, in the current context the code specifies adding some piece of evidence pro or contra some model selection.

The behaviour of watches is programmed in a dedicated KS. The KS is triggered by unit clause–objects representing atomic facts. When selected for execution, the KS **fetches** all the watches that possibly match the pattern of the fact. Thanks to the discrimination tree, this operation is very quick. Next, each of the retrieved watches is checked to see whether its pattern actually matches that of the fact. If this is the case, the fact is added to the list of matching facts of the watch, unless it belongs to the list already. (The **facts** attribute of a watch is an example of a dynamic object attribute; in contrast to **pattern** and **demons** attributes, which are fixed, **facts** is created and modified by the system.)

The next thing to do is to notify all the demons that the watch mentions the appearance of a new fact to match the watch's pattern. This point deserves a more detailed discussion. How can a demon be notified about such an event? At first sight it would seem that the solution is obvious. Adding a fact to the **facts** list of a watch makes an event. Now a KS should be written to be triggered by modified watches with non-empty **facts** lists. The KS could take the first fact (new elements are **consed** to the front of a list) from the list, and iterating through the demon list, 'implant' it in each of them, using some appropriate attribute. Modified in this way demons would consitute subsequent events, to be interpreted by another KS.

This straightforward solution does not take the non-determinism of the blackboard architecture into account. The causal chain depicted above of watch–demon activities would be interrupted by concurrently occurring events. Firstly, before the event meaning that a new fact was 'caught' by a watch could be processed, the watch could catch the next fact. Two events merge into one and the clarity of what really happened and what to do next is gone. Secondy and similarly, the notification of a demon may not have been processed before a new fact arrives from some watch.

The first problem we solve by introducing a new object type: **message**. The KS that checks whether new facts match any watches creates for each demon (from the demon list of a watch) a separate message stating the watch as a sender, the demon as an addressee, and the fact as data. The watch modification as such triggers no KS, but of course a new KS had to be written to process the messages.

The second problem, of a demon processing the incoming facts one at a time, is similarly solved by the above introduction of messages together with appropriate usage of the precondition of the message-processing KS.

The KS does not make a demon process a new fact directly; it tries to set a designated attribute **watch_fact** of the demon to the fact — the demon modified in this way will be processed as an event by still another KS. The point is that the KS for modifying a demon insists in its precondition that **watch_fact** not be set. In this way several KSARs can wait for the 'right' state of the demon. After having been processed the watch_fact is removed and the demon is ready to receive the next fact.

7. EXAMPLE CONSULTATION

To illustate the functioning of the program an example consultation with CONMOD is reproduced below. (Question topics are printed in boldface and options selected by the user indicated by arrows; for sake of compactness some options are left out. Normally, the user can 'escape' any question, which amounts to answering an unknown value.)

general field of application
→ structures
   machine elements
   metal forming
   geotechnical engineering
   biomechanics

problem type
   purely mechanical

→ coupled
purely thermal
other

desired level of sophistication of the analysis
→ simplified analysis (traditional design)
sophisticated assessment of the process

philosphy of analysis
→ complete analysis
. . .
. . .

coupled problem type
→ thermomechanical
fluid–structure interaction

problem dimensionality
   1D
      no bending
      bending
→ 2D
   → no bending
      → thin plate loaded 'in-plane'
         plane strain
         axisymmetric
     bending
   3D

expected magnitude of displacements
→ small
large

external agencies
→ static
   → deterministic
     probabilistic
   dynamic
     deterministic
     probabilistic

type of material
→ metal
polymer
timber
concrete
reinforced concrete

other multiphase (composite)
soil
rock
biomaterial

material symmetries
→ isotropic
   orthotropic
   anisotropic

Results of the consultation
Your problem should be posed as

$$\boxed{\text{Plane stress isotropic thermoelasticity}}$$

with the constitutive equation of the form:

$$\boldsymbol{\sigma} = \mathbf{C}[\boldsymbol{\varepsilon} - \alpha \mathbf{1}(\theta - \theta_0)]$$
$$\boldsymbol{\sigma} = \{\sigma_{xx}\ \sigma_{yy}\ \sigma_{xy}\}$$
$$\boldsymbol{\varepsilon} = \{\varepsilon_{xx}\ \varepsilon_{yy}\ \gamma_{xy}\}$$
$$\mathbf{1} = \{1\ \ 1\ \ 0\}$$

$$\mathbf{C} = \frac{E}{1-\nu^2}\begin{bmatrix} 1 & \nu & 0 \\ \nu & 1 & 0 \\ 0 & 0 & \frac{1-\nu}{2} \end{bmatrix}$$

References:
⋮

Remarks:
⋮

8. CONCLUSIONS

This chapter has discussed a method of embedding deduction in a blackboard system for constitutive modelling. The architecture provides powerful means of controlling inference. In particular, it can support resolution and combine it with other forms of computation.

We find expert systems technology of today to be excellent tool to deal with reasoning in the domain of constitutive modelling. The domain is practically very important and seemingly lends itself to formalization. Work in this area will certainly be important to those concerned with intelligent (and otherwise) materials databases and front-ends to numeric packages. The domain gives us also superb research opportunities in qualitative physics for future work.

The proposed method of combining the blackboard architecture with the resolution rule performs very well although redundant clauses are still generated. Implementing resolution requires that a mapping from resolution to blackboard concepts be developed. The mapping can be summarized as follows:

clause	→ clause–object
new clause derivation	→ event
suspended inference step	→ KSAR pointing to a resolve pair
inference strategy	→ blackboard scheduling rules and tests of KSs

The method rests on two main points:

(1) representing clauses as structured objects, and
(2) implementing resolution strategies in a general framework of blackboard activity scheduling.

Pattern–object notation is a convenient form for expressing predicate calculus formulas. In the notation the set of atomic formulas of a given sentence is separated from the logical structure of the sentence. Domain-dependent terms occur only in patterns of atomic formulas. The patterns are used to index clause–objects and in unification pattern matching.

Coupling resolution with the blackboard system seems very promising, but much work is needed to exploit fully the possibilities that begin to emerge.

It proves also very exciting to devise new types of objects that, for example, can carry procedurally encoded knowledge with them, and to provide a procedural semantics for such objects by writing accompanying knowledge sources. The system can actually be incrementally developed practically without writing new C code.

At present CONMOD finds itself in a considerable state of development. We have implemented reasoning paths leading to several constitutive models. Now we concentrate on widening the knowledge base. In the near future, we wish to incorporate some scheme for uncertainty handling, and to provide the system with an iconic user interface, which at the moment is rudimentary.

ACKNOWLEDGEMENT

The insightful assistance of Professor Z. Mróz at early stages of the project is gratefully acknowledged.

REFERENCES

[1] Ambroziak, J. R. and Kleiber, M., Qualitative reasoning about constitutive models, *Proceedings of the International Conference On Computational Engineering Science*, Atlanta, GA, 1988, Springer, New York, 1988.

[2] Bushnell, M. L. and Haren, P., *Blackboard Systems for AI: Special Issue*, in *Artif. Intell. Eng.*, **1**(2), 1986.

[3] Genesereth, M. R. and Nilsson, N. J., *Logical Foundations of Artificial Intelligence*, Morgan Kaufmann, Los Altos, CA, 1987.

[4] Hayes-Roth, B., A blackboard architecture for control, *Artif. Intell.*, **26**(3), 251–321, 1985.

[5] Nii, H. P., Blackboard systems, blackboard application systems, blackboard systems from a knowledge engineering perspective, *AI Mag.*, **6**(3), 82–106, 1986.

[6] Erman, L. D., London, P. E. and Fickas, S. F., The design and an example use of HEARSAY-III, *Proceedings of the Seventh International Joint Conference on Artificial Intelligence, Vancouver, BC, 1981*, William Kaufmann, Los Altos, CA, pp. 409–415, 1981.

[7] Lenat, D. B., Prakash, M. and Shepherd, M., CYC: using commonsense knowledge to overcome brittleness and knowledge acquisition bottlenecks, *AI Mag*, **6**(4), 65–85, 1986.

[8] deKleer, J., AMORD: a deductive procedure system, *Memo 435*, Artificial Intelligence Laboratory, Massachusetts Institute of Technology, Cambridge, MA, 1978.

[9] Charniak, E., Riesbeck, C. and McDermott, D., *Artificial Intelligence Programming*, Lawrence Erlbaum, Hillsdale, NJ, 1979.

[10] Doyle, J., A truth maintenance system, *Artif. Intell.*, **12**, 231–272, 1979.

[11] Doyle, J., A glimpse of truth maintenance, *Memo 461*, Artificial Intelligence Laboratory, Massachusetts Institute of Technology, Cambridge, MA, 1978.

[12] In depth: hypertext, Byte, **13**, 234–268, October 1988.

5

Interactive control of non-linear finite element calculations by an expert system

P. Wriggers
Universität Hannover
N. Tarnow
University of California at Berkeley

1. INTRODUCTION

The finite element method as a numerical approximation method to solve partial differential equations is at present the most powerful calculation tool in structural analysis. To use the far-reaching and manifold abilities of this method and to give the right interpretations of the obtained results requires not only extensive knowledge about the physics of the engineering task but also presupposes experience in dealing with the numerical algorithms of finite element programs. Because of the complexity of both domains, even well-versed users could fail in forming the appropriate models, choosing the necessary algorithms and setting the correct control parameters. The limited possibility of verifying the results often makes it difficult to notice these failures afterwards or to define the exact causes of poor results.

One step forward might be the integration of knowledge-based systems. Knowledge-based systems, which are often called expert systems (if the contents of the knowledge base justifies this expression), function as simulations of human problem-solving behaviour. These systems result from the artificial intelligence research of the last 20 years and, concerning practical usability, they belong to the most advanced products of this science. As controlling or supervising systems of a new quality, they could accompany a finite element computation from pre- to postprocessing and, in doing so, serve the user as well as the FE program as a competent adviser. The new quality is the possible processing of non-algorithmic and heuristic knowledge [1]. Expert systems could be applied to all different levels of modelling: from the real system via the physical model and the mathematical model up to the numerical computation model. They could be helpful in narrowing solution spaces and thereby lead to more reliable and expressive results (Fig. 5.1).

Former attempts to use expert systems in structural analysis aimed at advice for the user before the execution of a finite element calculation [2–4]. Moreover, FE calculation and expert system were conceptually divided. By the coupling of the expert system GENIUS and the FE program PCFEAP, this division has been closed [5]. The algorithmic calculation process is interactively accompanied by a knowledge-based system.

First, section 2 of this chapter gives an introduction to the applied FE program PCFEAP and to the expert system shell GENIUS, which has been developed in order to perform this study. The interaction of the two programs is shown in section 3. Finally, section 4 describes the applicability by two examples of finite element calculations.

2. THE FINITE ELEMENT METHOD PROGRAM PCFEAP AND THE EXPERT SYSTEM SHELL GENIUS

In this section, we should like to discuss the expert system GENIUS which has been developed at the Institut für Baumechanik und Numerische Mechanik of the University of Hannover and the analysis tool, the finite element program PCFEAP, developed by R. L. Taylor at the University of California, Berkeley.

2.1 The expert system shell GENIUS

2.1.1 General remarks

Expert system shells contain, apart from the knowledge base, all the components of an expert system. By adding the knowledge base, one could use an expert system shell to build an expert system for any domain. The expert system shell GENIUS uses conventional methods of artificial intelligence research and its conception adopts patterns from rule-based expert systems such as MYCIN, which was developed st Stanford University during the 1970s in order to diagnose certain bacteriological diseases [6–12]. The limited development duration has required some restrictions. For instance, no variables have been installed, so that unification algorithms could not be implemented.

2.1.2 Structure of the expert system shell GENIUS

Fig. 5.2 shows the basic structure of an expert system. The expert builds up a knowledge base with the help of a knowledge acquisition facility. The user communicates via an input/output interface with the inmference engine, which has the task of chaining the knowledge of the knowledge base and of drawing the correct conclusions from it. The explanation facility should help to make the processing of the expert system more transparent, in order to eliminate the black-box character which such a system might have.

GENIUS realizes the different components of an expert system, whose functions and interactions will be explained in this section, by various subroutines and files.

Ch. 5] CONTROL OF NON-LINEAR FINITE ELEMENT CALCULATIONS 99

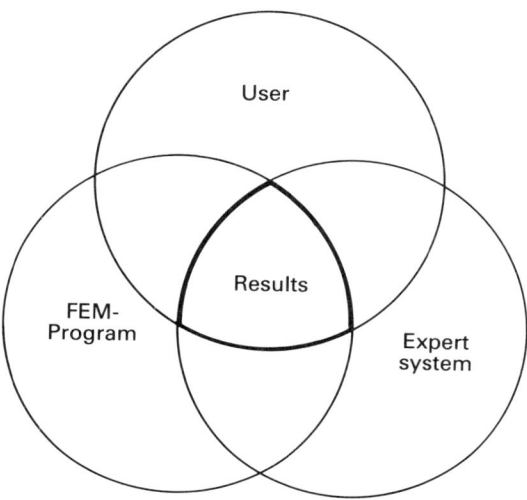

Fig. 5.1 — Improvement of numerical analysis by expert systems.

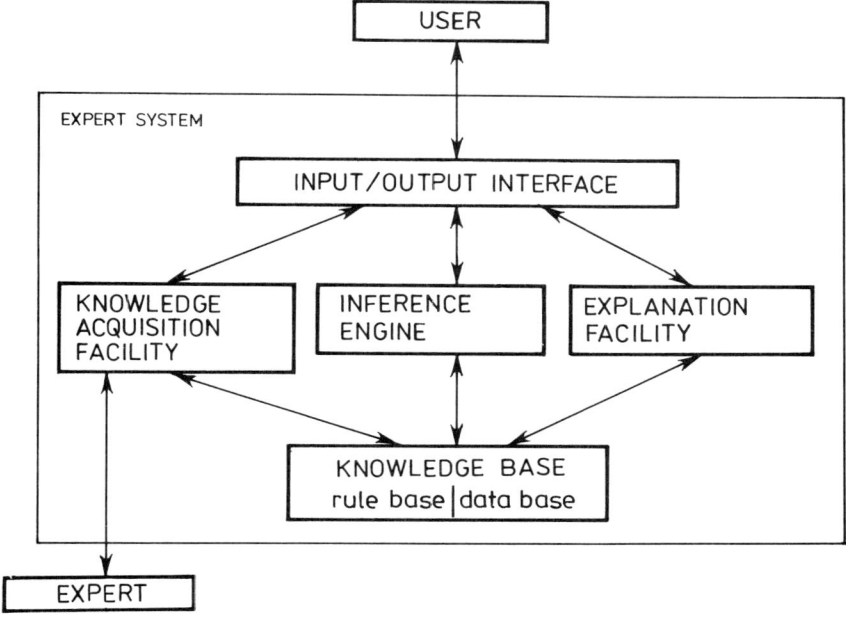

Fig. 5.2 — Modules of an expert system.

2.1.3 Knowledge representation in GENIUS

2.1.3.1 General remarks

The way of representing knowledge in the knowledge base is of fundamental importance for the nature and the ability of the entire system. The knowledge base is divided into a rule base, which contains the entire expert knowledge in form of rules, and a database, which stores all specific data about the current problem.

2.1.3.2 Rules

GENIUS distinguishes between two kind of rules: production rules, which contain the object knowledge, and meta-rules, which provide knowledge about the production rules and their arrangement.

A production rule is an independent incremental piece of knowledge of the following form:

$$premise \Rightarrow conclusion$$

Premise and conclusion consist of character strings, which are linked by the Boolean operators AND, OR and NOT. An example for a production rule might be:

```
RULE 23
IF
    the structure consists of straight bars
AND
    the bars are frictionless connected in nodes
AND
    forces act only on nodes
THEN
    the structure is an ideal truss
confidence factor: 100
```

Production rules connect objects or, more precisely, the statements of the premise with those of the conclusions to form so-called **chunks**. If the same statements occur in premises in conclusions of different rules, the chunks are connected to **clusters**, a sort of tree structure, which is also called a **semantic net**. Production rules allow an uncomplicated modelling of human expert knowledge; in addition, they have a structure which is easily processed by the computer. They are used for knowledge representation in nearly every commercial system.

Meta-rules help to reduce the search spaces and to increase the effectiveness of the system. In GENIUS this task is realized in the following way: meta-rules perform a dynamic management of the rule base by distinguishing betweeen an active and a passive part of the rule base. For this purpose, the rule base is divided into parts, each one containing knowledge about a certain problem's domain. If during a consultation the system concludes to a catchword, one of which is attached to each of the problem domains, the corresponding section is linked to the active part of the rule base. Thus, depending on the concluded data, the entire rule base is reduced to the relevant parts of it.

2.1.3.3 Knowledge acquisition

The extension of the rule base, and therefore an increase of knowledge, is easily carried out, apart from consultations, with the help of a text editor. This method of knowledge acquisition is called a passive learning ability. The rules are written down in a file, one beneath each other according to the example in section 2.1.3.2. To form rules, eight language elements are available: IF, IF NOT, THEN, THEN NOT, AND, AND NOT, OR and OR NOT.

During the development of a rule base, one must pay attention to avoid **dead** knowledge, which is isolated and without any connection to the rest of the rule base. Also, one has to ensure that no circular conclusions are formed, which keep the program captive in an endless loop, and that no contradictions in the rules occur.

2.1.3.4 Representation of uncertain knowledge

GENIUS distinguishes two sources of uncertainty: first, the limited validity of the rules and, second, the uncertainty with which the user answers questions of the system.

As a measure of evidence of the production rule itself, a confidence factor is attached to each rule, as shown in rule 23. This confidence factor ranges from 0, for a rule which is completely wrong, to 100, for a rule which is true in any case. Thus the concept of confidence factors allows the processing of vague knowledge and heuristics.

The answers of the users can be valued by them with one of the discrete rates: 0, 25, 50, 75, and 100.

2.1.4 Knowledge processing in GENIUS

2.1.4.1 Inference

The mechanisms which allow new knowledge to be attained and new conclusions to be arrived at from the rules stored in the knowledge base are called inference mechanisms. The inference strategy used in GENIUS is the so-called *modus ponens*. This says: if a statement A is known to be true and there is the rule 'if A, then B', then one can infer that statement B is also true. Although the implication between A and B permits only conclusions in one direction, we will see in the next section that one can use the *modus ponens* in both directions, in a forward and in a backward chaining manner, to infer new knowledge. An example for the application of the *modus ponens* might be:

Old verified knowledge: The temperature is over 500 degrees.

Rule: If the temperature is over 500 degrees, the material will melt.

New knowledge: The material is melting.

2.1.4.2 Processing control
The processing control merely has to master two tasks:

(1) It must be in position to start the inference process.
(2) It must decide which of the rules that are ready to start will be processed next and whether the chaining of the rules should be carried out forward or backward.

The first task is solved in this way: GENIUS searches in the database for rules which contain in their premises those statements that are given from the finite element program as described in section 3. These rules represent the start nodes of the semantic net, where the inference process begins.

The strategy used to solve the second task contains the intelligence of the system. GENIUS is in a position to perform forward chaining as well as backward chaining of the rules. The use of if–then rules and the use of the *modus ponens* lead directly to a forward chaining algorithm:

(i) A verified conclusion is given.
(ii) Rules where this conclusion appears in the premise are looked for.
(iii) In accordance with the *modus ponens* new conclusions can be drawn. The procedure starts again with (i).

Systems which exclusively use this method are called premise-controlled systems. However, as a mixed system GENIUS is also able to work in a conclusion-controlled manner:

(i) An assumption A is made.
(ii) All rules where A or not A appears in the conclusion are checked.
(iii) If the premises of these rules are fulfilled, A will be confirmed or disproved. In the case that it is unknown, if the premises are fulfilled, they will be used as new assumptions and the process will start with (i) again until the premises for the confirmation of refutation of A are finally cleared up.

Forward and backward chaining work hand in hand. GENIUS tries first to work with the given knowledge in a forward chaining way. If the user is not able to answer the system's questions which may occur during this process, GENIUS switches to backward chaining. GENIUS asks the user questions about the validity of not verified statements which are conjunctively connected with already verified statements, or when disjunctively connected statements in a conclusion demand the user's choice.

To make the decision as to which of several rules that may be ready to be processed is examined next, GENIUS follows the principle of the depth-first search. This means that GENIUS prefers a search in depth to a search in breadth. For this purpose, all obtained conclusions are written in a sequence which is worked up according to the principle of last in–first out. In comparison with the breadth-first search, this leads to a more natural question–answer process. A supposition is proofed first, before another is examined.

2.1.4.3 Processing of uncertain knowledge
The connection of confidence factors and information of the user allows the reliability of the obtained statements to be evaluated. During a consultation, the evidence of a certain statement may increase as well as decrease or even change into evidence for the counterstatement. The use of confidence factors in GENIUS does not allow a calculation of probabilities for instance in accordance to the proposition of Bayes. Nevertheless, the calculated certainties are an important means of comparing different options and of raising the efficiency of the system by stopping the inference for those statements whose confidence factors fall under a certain limit.

2.1.5 Explanation facility in GENIUS
The explanation facility helps to make the system more transparent and to remove the 'black-box character' in order to raise the user's acceptance.

At any question of GENIUS the user is offered the opportunity to switch to an information mode, which provides explanations of the, usually short, formulated questions or explains their relevance.

In addition, every decision is shown on the screen and commented. If a decision can not be made unequivocal, the user has the possibility to correct the system's interpretation.

The explanation facility uses an information base for its task, which can easily be created with an editor. In this file, one can write comments of any length to explain a certain rule. The connection with the rules of the knowledge base ensues from the rule numbers in the headlines of these comments.

2.2 The finite element method program PCFEAP
Since the solution of non-linear problems in most engineering applications is not given by analytical formulas, numerical methods have to be used. Within this context, the finite element method has been proved to be a flexible and powerful tool. Thus, for a general analytical part of an expert system, a finite element program is appropriate. Here we had to look for FE programs which could be combined with knowledge-based systems. Our choice was the highly modular FE program PCFEAP, developed by R. L. Taylor at the University of California at Berkeley [5], in which the algorithmic part is controlled by a macro language. Fig. 5.3 shows its basic program structure.

Several different elements and algorithms allow the application of the program to a wide variety of structural problems. This variety starts with linear problems and ends with fully non-linear — geometrically and physically — applications. Within this range, the user has a choice between different algorithms for the solution of his specific analysis.

Using finite elements the solution of a problem is performed in three steps. In the input phase, the problem data and the finite element mesh are specified. During the solution phase, the user selects the appropriate algorithms, which is done in the case of PCFEAP by a macro language. Finally, in the postprocessing phase, plots of the deformed geometry or the stresses are produced. Here, the user can choose between batch and interactive mode in the solution and postprocessing phases. In the following, two characteristic macro command sets for a linear and a non-linear application are shown:

Macro-commands in PCFEAP

Linear problem

assembly of system of equations $\mathbf{Kv}=\mathbf{f}$:	**tang,,1**
solution of equations :	**solv**
computation and printing of stresses :	**stre**
printing of displacements :	**disp**

Non-linear problem

loop over n_1 load steps :	**loop,,n1**
incrementation of load :	**time**
iteration loop :	**loop,,n2**
assembly of incremental system of equations $\mathbf{K}_T \Delta\mathbf{v} = -\mathbf{G}$:	**tang,,1**
solution of equations :	**solv**
end of iteration loop :	**next**
computation and printing of stresses :	**stre**
printing of displacements :	**disp**
end of loop over load steps :	**next**

As a result of the presence of the macro command language, the program PCFEAP is very well suited for a coupling with expert systems as will be discussed in the next section.

3. THE COUPLING OF GENIUS AND PCFEAP

Because PCFEAP provides numerical data while GENIUS performs symbol processing it was necessary to convert the quantitative data of the FE program into qualitative statements for the expert system. For this purpose, a quantitative–qualitative transformer was developed which uses a multitude of qualitative statements to describe the data of the numerical process. Appropriate statements are transferred to a database for later use in the expert system. In order to avoid preliminary conclusions, these statements must have a descriptive character rather than an evaluating character. Thus all explicit conclusions are performed in the expert system.

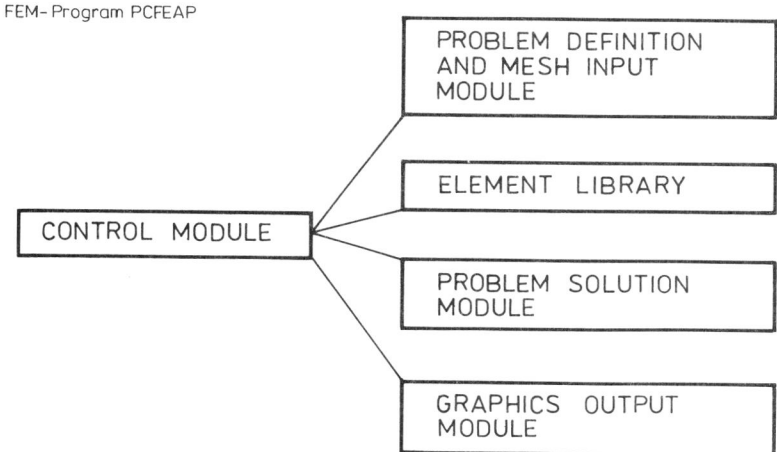

Fig. 5.3 — Modules of PCFEAP.

To increase the efficiency of the expert system, it is not useful to take every filed statement as a starting point for the inference process. Therefore only some of them are written into the queue which is processed during the consultation, while the rest serve as background information. The decision as to which of the two categories a statement belongs is fixed in the transformer and depends basically on whether the statement points to an anomalous computation or not. For example, the statement 'Number of negative pivot elements increased' would be relevant in this sense because it may indicate a change from a stable to an unstable state of the structure. However, the statement '6th residual norm calculated' can only be used as background information, for instance, during an examination of the convergence behaviour and in connection with a reported statement 'residual norm is increasing', which is again a statement of the first quality.

The transformation of the expert system's result into a usable form for the FE program is quite easy in the case of PCFEAP. As mentioned, PCFEAP is controlled by macro commands, which are symbols themselves. These commands can be transmitted easily to the FE program via a command module.

The quantitative–qualitative transformer represents a kind of input/output interface between the user, in this case the FE program, and the expert system. The schematic structure of the connection between PCFEAP and GENIUS is shown in Fig. 5.4.

4. CONTROL OF SOLUTION ALGORITHMS IN FINITE ELEMENT ANALYSIS

4.1 General remarks

The coupling of GENIUS and PCFEAP has been applied to three different structural applications:

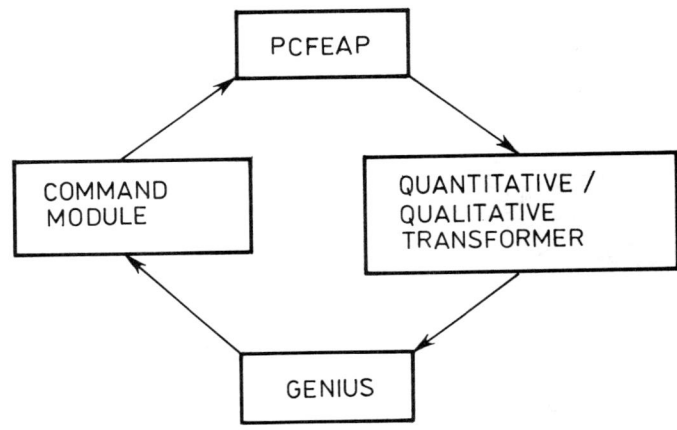

Fig. 5.4 — Coupling of PCFEAP and GENIUS.

(i) investigation of the convergence behaviour during Newtonian solutions of non-linear finite element equations;
(ii) control of elastic stability behaviour;
(iii) control of time steps for dynamic analysis based on Newmark's method.

The strategy in GENIUS is to detect, interpret and correct irregular solution paths and to give the user suggestions for changes in the algorithms. However, in regular solution paths GENIUS should not be noticed by the user. In cases where different corrections to a solution strategy are possible, GENIUS takes the one with the largest confidence factor.

GENIUS and PCFEAP were developed on an IBM-compatible personal computer using FORTRAN 77.

In the following two examples, a consultation of GENIUS within a finite element analysis is described. For this purpose, we first summarize the theoretical background of the application. Then we discuss the basic ideas of how to incorporate rules into the analysis and describe the qualitative–quantitative transformer. Finally, an example of a real consultation is presented.

4.2 Control of stability behaviour

The analysis of structural stability problems is very important in many fields of application such as beams and shells. However, if such a non-linear buckling problem is to be solved, account should be taken of the many different factors that influence the behaviour of the solution and with this the choice of the appropriate algorithm. Thus an expert system may be well suited to guide a non-expert user through the finite element analysis of a stability problem. Some aspects of the development of an expert system for this problem class will be addressed in the following.

4.2.1 Stability problems in elastostatics
The basic phenomena associated with stability problems are depicted in Fig. 5.5. A pure stress problem is given by curve C_1 which shows a one-to-one

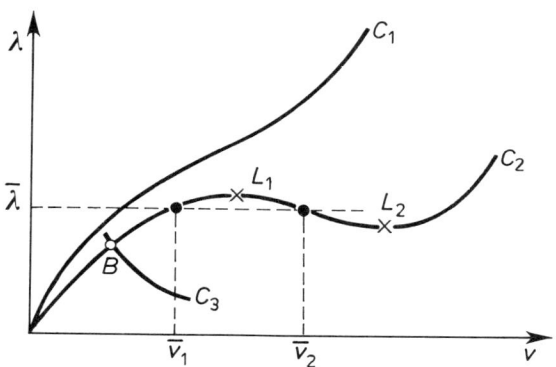

Fig. 5.5 — General non-linear structural behaviour.

correspondence between the displacement state and the load statre. In contrast to that, a stability problem exhibits different equilibrium states, $\bar{v}_1$, $\bar{v}_2$, for a given load level $\bar{\lambda}$ (see curve C_2). Stability theory makes a distinction between points where the load deflection curve exhibits maxima or minima, i.e. **limit points**, and points where a branching of the solution is possible, i.e. **bifurcation points**. In Fig. 5.5 limit points are denoted by L_1, L_2; the intersection of C_2 and C_3 depicts a bifurcation point B.

For our purpose, we do not need to specify the finite element model used in the calculation. Thus, we formulate the discrete form of the non-linear equilibrium equations which are given below in matrix notation:

$$\mathbf{G}(\mathbf{v},\lambda) = \mathbf{R}(\mathbf{v}) - \lambda\mathbf{P} = \mathbf{0} \tag{1}$$

Here $\mathbf{R}(\mathbf{v})\in\mathfrak{R}^n$ is the so-called stress divergence term which has to be equal to the loading $\lambda\mathbf{P}$ in the equilibrium point, n is the number of finite element equations used to describe the problem and λ is the load parameter.

Solution of eqn (1) by a Newton–Raphson procedure leads to the following incremental system of equations:

$$\mathbf{K}_T(\mathbf{v}^i)\, \Delta\mathbf{v}^{i+1} = -\mathbf{G}(\mathbf{v}^i, \lambda)$$
$$\mathbf{v}^{i+1} = \mathbf{v}^i + \Delta\mathbf{v}^{i+1} \qquad (2)$$

where the tangential stiffness matrix $\mathbf{K}_T$ is given by the directional derivative of eqn (1) with respect to the displacements $\mathbf{v}$.

In a stability point, $\mathbf{K}_T$ becomes singular. This can be described (see for example [14]) by

$$\det \mathbf{K}_T = 0 \qquad (3a)$$
$$(\mathbf{K}_T - \omega_i \mathbf{1})\, \boldsymbol{\phi}_i = \mathbf{0} \quad \text{with} \quad \omega_i = 0 \qquad (3b)$$

In eqn (3b) ω_i denotes the ith eigenvalue and $\boldsymbol{\phi}_i$ the ith eigenvector of $\mathbf{K}_T$. A non-trivial solution of eqn (3b) is equivalent to eqn (3a) which can be used to trace stability points on a non-linear solution curve. The determinant in eqn (3a) can be simply computed from the diagonal matrix $\mathbf{D}$ in the standard triangularization $\mathbf{K}_T = \mathbf{L}\mathbf{D}\mathbf{L}^T$, which leads to the following formula:

$$\det \mathbf{K}_T = \prod_{i=1}^{n} D_{ii} \qquad (4)$$

where D_{ii} are the diagonal elements of $\mathbf{D}$. A classification of equilibrium states is then given by

$$D_{ii} > 0 \rightarrow \mathbf{K}_T \text{ is positive definite} \rightarrow \text{stable equilibrium}$$
$$D_{ii} = 0 \rightarrow \mathbf{K}_T \text{ is semidefinite} \qquad \rightarrow \text{indifferent equilibrium}$$
$$D_{ii} < 0 \rightarrow \mathbf{K}_T \text{ is indefinite} \qquad \rightarrow \text{unstable equilibrium}$$

The eigenvectors associated with zero eigenvalues denote the snap-through or buckling modes. The latter can be used to perform a branch switching into secondary paths of the global solution [13]. For this purpose, the following formula can be used:

$$\bar{\mathbf{v}} = \mathbf{v} + \xi \frac{\boldsymbol{\phi}_i}{\|\boldsymbol{\phi}_i\|} \qquad (5)$$

Thus a perturbation of the displacement field with the scaled buckling mode leads to branch switching. However, in this formula the factor ξ is unknown and has to be estimated by heuristic methods or formulas.

4.2.2 Basic considerations for the use of GENIUS

An expert system for stability problems should be able to determine and identify critical points such as bifurcations and limit points. Furthermore, it should provide the user with the capability of an automatic branch switch into secondary paths.

GENIUS monitors the change of the determinant of $\mathbf{K}_T$ which gives some information about the possibility of an approaching singularity. An indicator for such a critical point is a change in the number of negative diagonal elements in $\mathbf{K}_T$, which is easily obtained within the triangularization process. Thus, if such a negative diagonal element occurs, one knows that a stability point has been passed by the solution algorithm. For the exact location of the stability point, we use a bisection method as described in [13] which is based on an arc length scheme. The iteration is stopped when the absolute value of the eigenvalue in eqn (3b) satisfies the inequality $|\omega_0| < 10^{-5}$. The appearance of multiple eigenvalues is also indicated by GENIUS. After the critical point is located within a given tolerance, we have to distinguish between limit and bifurcation points. For this purpose, we need knowledge about the buckling mode ϕ associated with this stability point. The following formulas hold:

$$\phi_i^T \mathbf{P} \neq 0 \Rightarrow \text{limit point}$$
$$\phi_i^T \mathbf{P} = 0 \Rightarrow \text{bifurcation point} \quad (6)$$

In case of a bifurcation point, the user may wish to follow the secondary solution path. Then branch switching has to be performed. A successful branch switch depends on a good choice of the factor ξ in eqn (5). The following heuristic formula leads in many applications to a good estimate for ξ:

$$\xi = \frac{\|\mathbf{v}\|}{(\mathbf{v}^T \phi_0 / \|\mathbf{v}\| \|\phi_0\|) \times 100 + 1} \quad (7)$$

Here, ϕ_0 is the eigenvector associated with ω_0 and $\mathbf{v}$ denotes the actual displacement vector.

GENIUS has to check whether the branch switch was successful or not. Criteria for a successful switch are:

(i) a smaller number of negative diagonal elements in the case of a positive slope of the secondary solution curve;
(ii) a negative slope of the secondary path.

It should be noted that the switch to a secondary path takes several iterations steps. After these steps GENIUS informs the user about the state of the solution.

Remarks
The iteration procedure to compute singular points, the orthogonality check (6) and the computation of ξ are performed in the command modul within PCFEAP. Thus, one cannot construct an expert system for the task described above which has no interface to a conventional finite element program. Let us remark that the questions — e.g. about the orthogonality between **P** and ϕ — which GENIUS addresses to PCFEAP can be viewed as a process of back chaining in the sense of the definition given in section 2.1.4.2.

4.2.3 Quantitative–qualitative transformer
When GENIUS monitors the determinant the quantitative–qualitative transformer provides the user near a singular point with the following message

    ABS(DET/DET-0)<0.1
    ABS(DETERMINANT) DECREASES

When a singular point is passed or when the solution path is unstable the following message is given:

    NUMBER OF NEGATIVE PIVOTS INCREASES
    NEGATIVE PIVOT ELEMENTS EXIST

After the analysis of a singular point the messages

    BUCKLING MODE ORTHOGONAL TO LOADVECTOR
    MORE THAN ONE ZERO EIGENVALUES EXIST

indicate that the singular point is a multiple bifurcation point.

A successful branch switch is related to the next two messages:

    LOADPARAMETER DECREASES AFTER BRANCH-SWITCH
    NUM. OF NEG. PIVOTS DECREASES AFTER
    BRANCH-SWITCH

All these messages will be filed into the waiting queue of the inference engine.

4.2.4 A consultation example with GENIUS
In this section, the truss system depicted in Fig. 5.6 is investigated. It is assumed that a single truss member can buckle as well as the overall structure. Thus, despite its simplicity, the load–deflection curve in Fig. 5.7 exhibits limit and bifurcation points.

In Fig. 5.7 the determinant of $\mathbf{K}_T$ is plotted versus the mid-displacement of the truss structure.

The computation of the response is started from load level zero with a

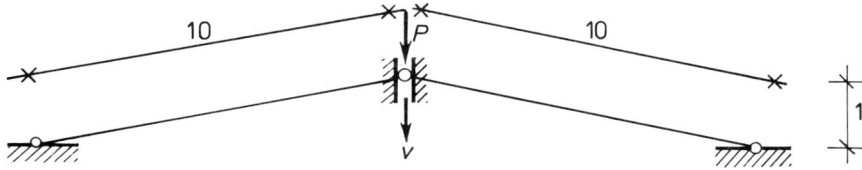

Fig. 5.6 — Truss structure.

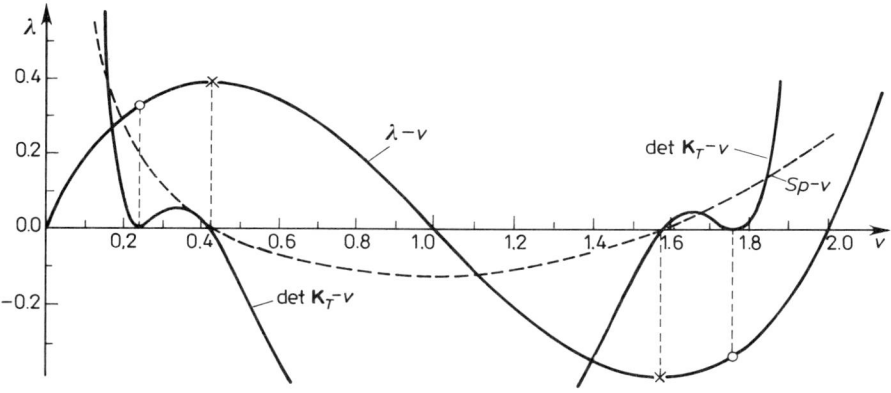

Fig. 5.7 — Load–deflection curve.

given load increment. In this example the macro commands in PCFEAP are chosen such that a pure Newtonian method is applied; see macro commands in section 2.3. The decrease of the determinant during the successive incrementation of the load yields the following message:

   APPROACHING INSTABILTY   with 100.0%

Next a singular point on the load deflection curve is passed. This leads to the information

   SYSTEM IS IN UNSTABLE CONDITION   with 100.0%

Now GENIUS concludes

   SINGULAR POINT DISCOVERED   with 100.0%
   "GENIUS" ARRANGES: back,oegv   with 100.0%

  The macro command **back,oegv** forces PCFEAP to switch to the arc length scheme and to compute the singular point via the bisection method.

This leads to the computation of the first zero eigenvalue which is shown in Fig. 5.8.

An investigation of formulas (6), using the corresponding eigenvector in PCFEAP, yields the following message in GENIUS:

>
> INSTABILITY IS A BIFURCATION
> AND
> DO YOU WANT TO STAY ON PRIMARY BRANCH   F1
> OR
> YOU WANT TO SWITCH TO A SECONDARY BRANCH
> AND
> "GENIUS" ARRANGES: time,add   F2
> OR
> COMPUTATION SHOULD BE STOPPED   F3
> OR
> YOU CANNOT TAKE A DECISION   F4
> OR
> NO OPTION APPLIES TO THE PROBLEM   F5

We assume that the user wants to branch into the secondary path and therefore types F2. Here, the macro commands **time,add** mean that PCFEAP will use eqn (5) to add the buckling mode to the displacement field where eqn (7) is applied for scaling. During the iterations for the switch GENIUS informs the user that PCFEAP is still in the iteration process and that the structure is unstable. Finally rule 29 is fulfilled:

>
> RULE 29
> IF
>   loadparameter decreases
> OR
>   num. of neg. pivots decreases
> THEN
>   switch to secondary path successful
> confidence factor 100.

and the following message appears:

>
> SWITCH TO SECONDARY PATH SUCCESSFUL   with 100.0%

The secondary branches are shown in Fig. 5.9.

GENIUS concludes that the secondary path is a stable branch. However, after some load increments a message appears which indicates a new singular point. The following investigation of the type of stability shows that a limit point occurred which is associated with the secondary branch (II.2). Now GENIUS provides the user with the menu:

Ch. 5] CONTROL OF NON-LINEAR FINITE ELEMENT CALCULATIONS 113

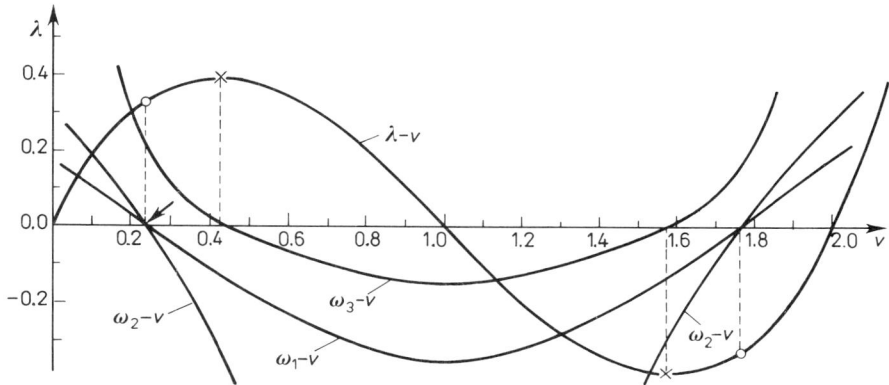

Fig. 5.8 — Eigenvalues of $\mathbf{K}_T$ as a function of the mid-displacement.

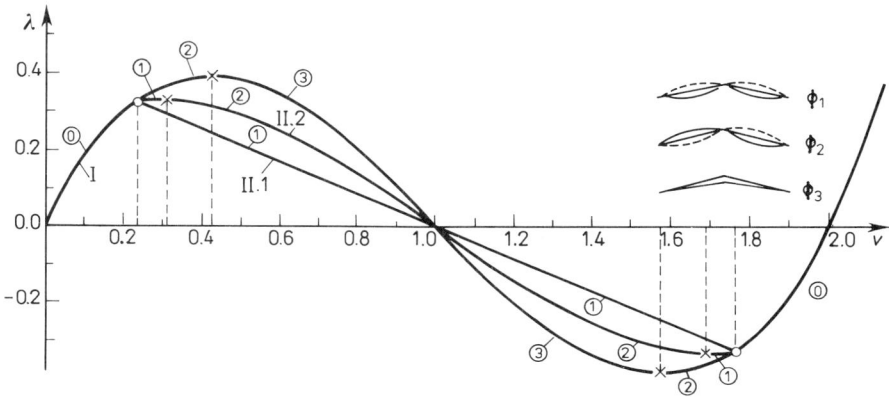

Fig. 5.9 — Secondary equilibrium paths of the truss structure.

INSTABILITY IS A LIMIT POINT	
AND	
YOU WANT TO CARRY ON	F1
OR	
YOU WANT TO QUIT	F2
OR	
YOU CANNOT TAKE A DECISION	F3
OR	
NO OPTION APPLIES TO THE PROBLEM	F4

If the user wants to proceed with the calculations, he may type F2 and PCFEAP will follow the secondary path. During these iterations, GENIUS will repeat the message that the solution path is unstable.

4.3 Control of the time step for dynamic analysis with the Newmark method

The use of numerical integration methods for the computation of time-dependent problems in finite element analysis requires an expert knowledge of how to choose an appropriate time step for the problem at hand. In this section the expert system GENIUS is used to provide an estimate for the time step which should on the one hand lead to an inexpensive analysis and on the other hand meet the requirements of accuracy.

4.3.1 The Newmark method

As a direct integration method the Newmark method is based on the idea of satisfying the dynamic equilibrium equations at discrete times:

$$\mathbf{M}_{t+\Delta t}\, \ddot{\mathbf{v}} + \mathbf{C}_{t+\Delta t}\, \dot{\mathbf{v}} + \mathbf{K}_{t+\Delta t}\, \mathbf{v} = \mathbf{P} \tag{8}$$

with the mass matrix $\mathbf{M}$, the damping matrix $\mathbf{C}$, the stiffness matrix $\mathbf{K}$ and the time-dependent load vector $\mathbf{P}$; $\mathbf{v}$ is the displacement vector of the set of finite element equations, $\dot{\mathbf{v}}$ denotes the velocities and $\ddot{\mathbf{v}}$ is the acceleration vector.

The following assumptions for velocities and displacements yield the standard implicit algorithm:

$$\begin{aligned}
\dot{\mathbf{v}}_{t+\Delta t} &= \dot{\mathbf{v}}_t + [(1-\delta)\, \ddot{\mathbf{v}}_t + \delta\, \ddot{\mathbf{v}}_{t+\Delta t}]\, \Delta t \\
\mathbf{v}_{t+\Delta t} &= \mathbf{v}_t + \dot{\mathbf{v}}_t \Delta t + [(\tfrac{1}{2} - \alpha)\, \ddot{\mathbf{v}}_t + \alpha\, \ddot{\mathbf{v}}_{t+\Delta t}]\, \Delta t^2
\end{aligned} \tag{9}$$

α and δ are parameters which have to be chosen such that the accuracy and stability of the method are ensured. For $\alpha = 0.25$ and $\delta = 0.5$, the method is unconditionally stable and shows optimal accuracy. However, the remaining integration error causes a period elongation which is a function of $\Delta t/T$, where Δt is the chosen time step and T a period of vibration.

4.3.2 Considerations for the use of GENIUS

GENIUS has the task of checking whether the chosen time step leads to a solution which meets the user's requirements concerning accuracy and effort. Moreover, GENIUS has to make sure that the finite element model fulfils the basic requirements for performing the dynamic analysis, e.g. to check whether the finite element mesh is able to reproduce all essential deformation patterns in its response. However, in this section only the considerations concerning the time step control will be explained in detail.

A first estimate for a time step used in an implicit time integration is given by (see for example Bathe [15])

$$10 \, \Delta t < T_s < 20 \, \Delta t \tag{10}$$

This choice is based on considerations concerning stability and period elongations of the integration scheme. In eqn (10) T_s corresponds to the smallest period in the system.

The smallest period T_s which has to be considered within the computation of the response of the structure depends on the structure, the frequency content and the distribution of the loading. The response ratio of a single-degree-of-freedom system shows (see for example Bathe [15]) that dynamic effects can be neglected for a ratio between the frequency of the loading and the eigenfrequency of the system smaller than 0.25. Thus we can conclude that, if the highest frequency of the loading is known, our finite element model should be able to respond in frequencies lower than or equal to one-fourth of this value. If we apply this conclusion to eqn (10), we obtain a relation between the time step Δt and the shortest period of the loading T_p:

$$40 \, \Delta t < T_p < 80 \, \Delta t \tag{11}$$

Furthermore, we can conclude that a dynamic analysis is not necessary if the lowest eigenfrequency of the system is higher than one-fourth of the highest frequency of the load spectrum. For example, the same wind loading which requires a dynamic analysis for a tower may only lead to static conditions for a windowpane, whose eigenfrequencies are much higher than those of the tower.

4.3.3 The quantitative–qualitative transformer

Various information which is necessary in order to evaluate the quality of the finite element model for dynamic analysis is given by the input file of PCFEAP; other information can be determined by using the macro command environment. However, there is no possibility of obtaining an analysis of the load history function directly from the finite element program. Therefore, an additional task of the quantitative–qualitative transformer in this case is to examine the loading with respect to the frequency contents and to present the results in a qualitative form. Since we are only interested in qualitative statements, a Fourier analysis would be too costly. Instead of this, the transformer performs a simple qualitative Fourier analysis. The loading history function which is given by the user is divided into a group of single rectangular impulses according Fig. 5.10.

This assumption is permissible since every function other than the single rectangular impulse would show a stronger convergence in the frequency domain. It can be shown that we cover about 90% of the impulse energy if the function of the rectangular impulse in the frequency domain is truncated after the first root ($2\pi/T$). Thus, for every rectangular impulse of length T, the frequency $1/T$ is the highest frequency and T is the shortest period of the impulse that we will consider (Fig. 5.11).

Finally, all rectangular impulses, which are characterized by their

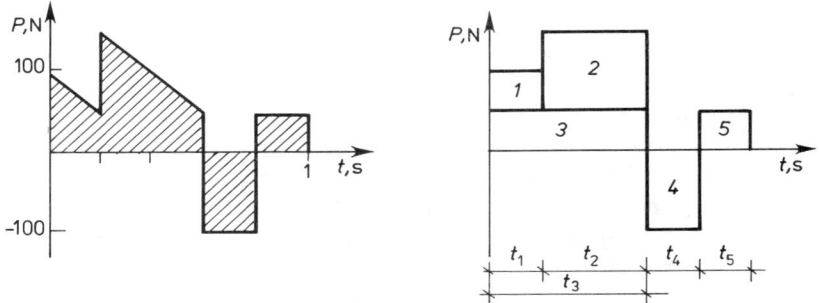

Fig. 5.10 — Example of load history and its division into rectangular impulses.

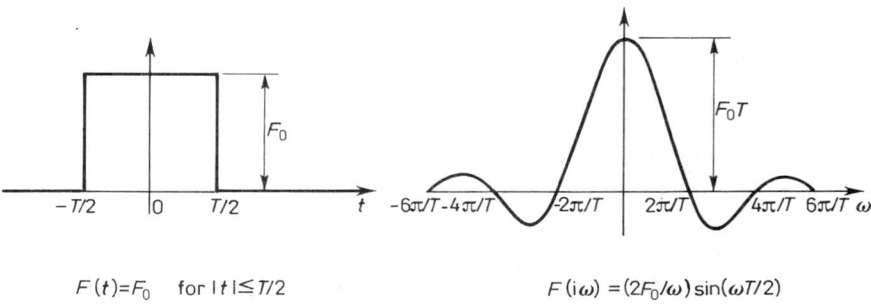

$F(t) = F_0$ for $|t| \leq T/2$ $F(i\omega) = (2F_0/\omega)\sin(\omega T/2)$

Fig. 5.11 — Rectangular impulse in time and frequency domain.

lengths and their proportions as a percentage of the entire load history function, are divided into the following five groups:

$$T < 20\,DT$$
$$20\,DT < T < 40\,DT$$
$$40\,DT < T < 60\,DT$$
$$60\,DT < T < 80\,DT$$
$$T < 80\,DT$$

Each group is connected to a number that represents its proportion as a percentage of the load history function which is covered by this particular interval. With this method we found a qualitative representation of the analysis of the loading. The intervals are used as statements for GENIUS. The attached number is interpreted as a confidence factor. In addition, the transformer provides a statement about the continuity of the load history function, where the confidence factor expresses the grade of discontinuity.

4.3.4 GENIUS in the consultation

The load history as shown in Fig. 5.10 may be regarded as a loading of a windowpane. The chosen time to step is 0.01 s. The quantitative–qualitative transformer files, among other, the following messages to the database of GENIUS:

T>80 · DT	.0
60 · DT<T<80 · DT	37.9
40 · DT<T<60 · DT	20.2
20 · DT<T<40 · DT	40.7
T<20 · DT	.0

GENIUS notices that it is consulted before the first time step of a dynamic calculation. Therefore it asks the user to choose one of the following options:

DYNAMIC ANALYSIS NECESSARY	F1
OR	
CHECK WHETHER STATIC ANALYSIS IS SUFFICIENT	
AND	
"GENIUS" ARRANGES: tang,mass,subs,expe	F2
OR	
YOU CANNOT TAKE A DECISION	F3
OR	
NO OPTION APPLIES TO THE PROBLEM	F4

In our case the user may not be sure whether a dynamic analysis really is necessary and selects F2. With the macro commands **tang,mass,subs,rxpe**, GENIUS initiates the following sequence of tasks in PCFEAP: computation of tangent stiffness (**tang**) and mass matrix (**mass**), calculation of the structure's smallest eigenfrequency (**subs**). The quantitative–qualitative transformer (**expe**) represents the result as follows:

1.	eigenfrequency checked	100.0
not 1.	eigenfrequency>4 exciter frequency	100.0

Here, **exciter frequency** means the reciprocal of the shortest duration which was found in the qualitative Fourier analysis. From this messages GENIUS concludes

GENIUS concludes from RULE 528:	
DYNAMIC ANALYSIS NECESSARY	with 100%

GENIUS tries next to obtain information about the user's requirements concerning the accuracy of the calculation:

ANALYSIS AS EXACT AS POSSIBLE	F1
OR	
REASONABLE ANALYSIS	F2
OR	
ONLY BASIC RESPONSE IN DEFORMATIONS	F3

```
                OR
                YOU CANNOT TAKE A DECISION                    F4
                OR
                NO OPTION APPLIES TO THE PROBLEM              F5
```

Again, the user is not sure and chooses **F4**, on which GENIUS switches to the backward chaining mode and asks

```
                How do you estimate the probability of
                THIS IS A DIMENSIONING PROBLEM
                0%                                             F1
                25%                                            F2
                50%                                            F3
                75%                                            F4
                100%                                           F5
                NO OPTION APPLIES                              F6
```

The answer **F5** stops backward chaining and yields:

```
                GENIUS concludes from RULE 523
                    NEED FOR A ECONOMICAL BUT GOOD ANALYSIS
                                                        with 100%
```

This message and the transformer's statement about the discontinuity of the load–time function are used now to reason about which ratio between the time step and the lengths of single rectangular impulses is appropriate:

```
        40 · DT<T<60 · DT recommendable                  57.5
        60 · DT<T<80 · DT recommendable                  43.3
```

The confidence factors in this case are not to be interpreted in such a way as to imply that 57.5% or 43.8% of the impulse durations should be in the appropriate interval. Instead, they indicate that some intervals are more recommendable for this specific problem than others. The final step is to compare these statements with the actual situation as reported by the transformer with rules such as the following and thus to obtain statements as to whether and how the time step should be changed:

```
        RULE 517
          IF
        20 · DT<T<40 · DT
          AND
        60 · DT<T<80 · DT recommendable
          THEN
        DT should be reduced
        Confidence factor 75.
```

It turns out that there is some evidence for a reduction of Δt:

```
        GENIUS concludes from RULE 517
            DT SHOULD BE REDUCED                      with 30.5%
```

Consequently, GENIUS reduces the time step by 50% and repeats the analysis. The reduced time step yields now a confidence factor of 23 for an increase. Since forward chaining is only continued for statements with a confidence factor of more than 25, the analysis stops here. The reduced time step is evaluated to be appropriate for our problem. If there were again sufficient evidence for a change of the time step, GENIUS would proceed by changing the time step again until an appropriate value was found.

REFERENCES

[1] Tarnow, N., Expertensysteme im Bauingenieurwesen, *Bauingenieur*, **10** 468, 1987.
[2] Bennet, J., Crealy, L., Englemore, R. and Melosh, R., SACON: a knowledge-based consultant for structural analysis, *Tech. Rep. STAN-CS*-78-699, Stanford University, 1978.
[3] Rivlin, J. M., Hsu, M. B. and Marcal, P. V. Knowledge based consultation for finite element analysis, *Tech. Rep. AFWAL-TR*-80-3069, MARC Analysis Research Corp., Palo Alto, CA, 1980.
[4] Taig, I. C., Expert aids to reliable use of finite element analysis. In *Reliability of Methods for Engineering Analysis*, Pineridge Press, Swansea, 1986.
[5] Taylor, R. L., PCFEAP — a finite element program for personal computers, *User Manual*, Department of Civil Engineering, University of California, Berkeley, 1987.
[6] Shortlife, E. H., *Computer Based Medical Consultations*: *MYCIN*, North-Holland, Amsterdam, 1976.
[7] Buchanan, B. and Shortlife, E. H., *Rule-Based Expert Systems*, Addison-Wesley, Reading, MA, 1984.
[8] Hayes-Roth, F., Waterman, D. E. and Lenat, D. B., *Building Expert Systems*, Addison-Wesley, Reading, MA, 1983.
[9] Level Five Research Inc., Insight knowledge system, version 1.2, *Manual*, Melbourne Beach, 1985.
[10] Hartmann, D. and Lehner, K., *Technische Expertensysteme*, Weiterbildendes Studium Bauingenieurwesen, Universität Hannover, 1987.
[11] Harmon, P. and King, D., *Expertensysteme in der Praxis*, Oldenbourg, München, 1986.
[12] Winston, P. H., *Artificial Intelligence*, Addison-Wesley, Reading, MA, 1984.
[13] Wagner, W. and Wriggers, P., A simple method for the calculation of postcritical branches, *Eng. Comput.* **5** 103–109, 1988.
[14] Wriggers, P., Wagner, W. and Miehe, C., A quadratically convergent procedure for the calculation of stability points in finite element analysis, *Comput. Methods Appl. Mech. Eng.*, **70** 327–349, 1988.
[15] Bathe, K. J., *Finite Element Procedures in Engineering Analysis*, Prentice-Hall, Englewood Cliffs, NJ, 1982.

6

Concepts for the application of AI techniques in computational mechanics

D. Hartmann
Ruhr Universität, Bochum

1. INTRODUCTION

One or two decades ago, within computational engineering no one paid attention to artificial intelligence (AI). AI was considered a research toy for sophisticated computer scientists. Reasonably, the development of computer tools for analysis purposes towards more maturity (e.g. finite element techniques) was of prime interest.

Since then, the scenario has dramatically changed. Because of the impact of the 'fifth-generation computer projects' all over the world, 'AI techniques' have become one of the *ne plus ultra* research fields. As a matter of fact, currently AI is incorporated into and/or coupled with conventional computer-aided engineering (CAE). In particular, knowledge- and rule-based systems have attracted a lot of attention. Without exaggeration, it can be stated that a new computer era has begun with regard to software and hardware.

2. CONVENTIONAL COMPUTATIONAL MECHANICS

It is well-known that the mathematical representation of mechanical phenomena is primarily based on ordinary and partial differential equations. Within recent years, in practice, problems have been solved by means of two main approaches. First, analytical mathematics has been applied to the solution of the individual differential equation. Second, the original problem has been transformed into a numerical equivalent through finite computational methods (e.g. finite differences, finite strips, finite element, boundary elements). Together with the computer evolution the second approach definitely became the dominant method of 'problem solution'.

3. NEW DIRECTIONS IN COMPUTATIONAL MECHANICS

Very recently, the advent of AI techniques in computer science led to new frontiers. To a certain extent, conventional methods are reviving, but in terms of a completely different philosophy. The aspect of general computer-assisted problem solving becomes more and more popular. Since problem solving mechanisms are associated with expertise and knowledge about a specific knowledge domain, knowledge-based computer systems (or so-called expert systems) have been created. Obviously, there is a tremendous need for expert systems because it is evident that a lot of real-world problems cannot solely be represented through number-crunching algorithms. The following areas can be identified as characteristic application fields for expert systems:

— problems associated with a very complex internal structure;
— problems for which definite theories and algorithms are not available;
— problems that incorporate a wide variety of diverse knowledge concentrated in the brains of many experts;
— problems that consist of an exponential or even a combinatorial multitude of candidate solutions.

In fact, the generic solution of an optional computational mechanics problem falls into the above category. If not only the algorithmic part of the solution is considered but also operational steps such as

— problem classification,
— selection of appropriate solution methods,
— problem modelling,
— verification,
— error diagnosis,
— process monitoring,
— result evaluation,
— tutoring of noices,
— training of experts,
— electronic user documentation, etc.,

the necessity of AI techniques in computational mechanics is evident. Of course, conventional numerical methods have to be linked to AI-based systems to create applicable tools for practical engineering tasks.

Within the last few years the significance of knowledge-based systems has been recognized. Thus a variety of software and AI computer languages (e.g. expert system shells, LISP, PROLOG, Smalltalk, etc.) is available that facilitates knowledge engineering, particularly for non-computer scientists. Undoubtedly, expert system technology is the first AI branch that has attained maturity.

4. MECHANICAL SOLUTION ASSISTANT

Within this contribution, the concept of a mechanical solution assistant for plane stress problems is demonstrated. The assistant is based on an expert system shell called INSIGHT [1] that runs on IBM-PC-compatible computer systems. Since a shell provides all the characteristic facilities that an expert system requires, except for the problem-specific knowledge base, it constitutes an excellent computer tool for prototyping complex knowledge-based systems [2]. In other words, using a shell frees the user from implementing otherwise needed facilities such as

— inference engine,
— explanation facility,
— user interface,
— knowledge acquisition facility.

The knowledge representation in INSIGHT is based on so-called production rules (PR) as a format for the elementary description of knowledge and information. The corresponding production rule language INSIGHT provides a syntax by means of which the logical flow of questions (in accordance with the underlying inference scheme) and conclusions can be presented to the user. As additional information and new knowledge are gained, new cognitive items can be added to the knowledge base (rudimentary modification facility). Reserved key words and operators (in capital letters) are utilized to program the knowledge base. Examples of reserved words that are more or less self-explanatory are given below:

AND	DISPLAY	IF	RULE
ARE	ELSE	IS	THEN
CF	END	OFF	THRESHOLD
CONFIDENCE	EXPAND	ON	TITLE

As a minimum, production rules have the following general form:

RULE	rule name
IF	supporting condition
THEN	conclusion

Of course, more complicated constructs may be specified, e.g. more than one supporting condition or multiple conclusions may be needed (AND keyword). Also, rules may be nested and related to other rules or grouped together. The key words EXPAND and DISPLAY provide a means for displaying text and simple graphics to improve the understandability of the reasoning process.

The specified rules are related to goals and subgoals. Goals or subgoal statements consist of a phrase or sentence which describes a conclusion that can be reached via the inference facility (a backward chaining strategy is used in INSIGHT). At least one goal is mandatory. Each goal must be preceded by a goal number and must correspond to some knowledge about the goal that can prove or reject it. To exemplify a goal, the following section of the mechanical solution assistant is given:

! Primary goals !
1. Problem classification identified
1.1 Finite element method (FEM)
1.2 Boundary element method (BEM)
1.3 Symbolic manipulation approach (SMA)

The above-indicated goals represent the main functionality of the expert system which is graphically shown in Fig. 6.1.

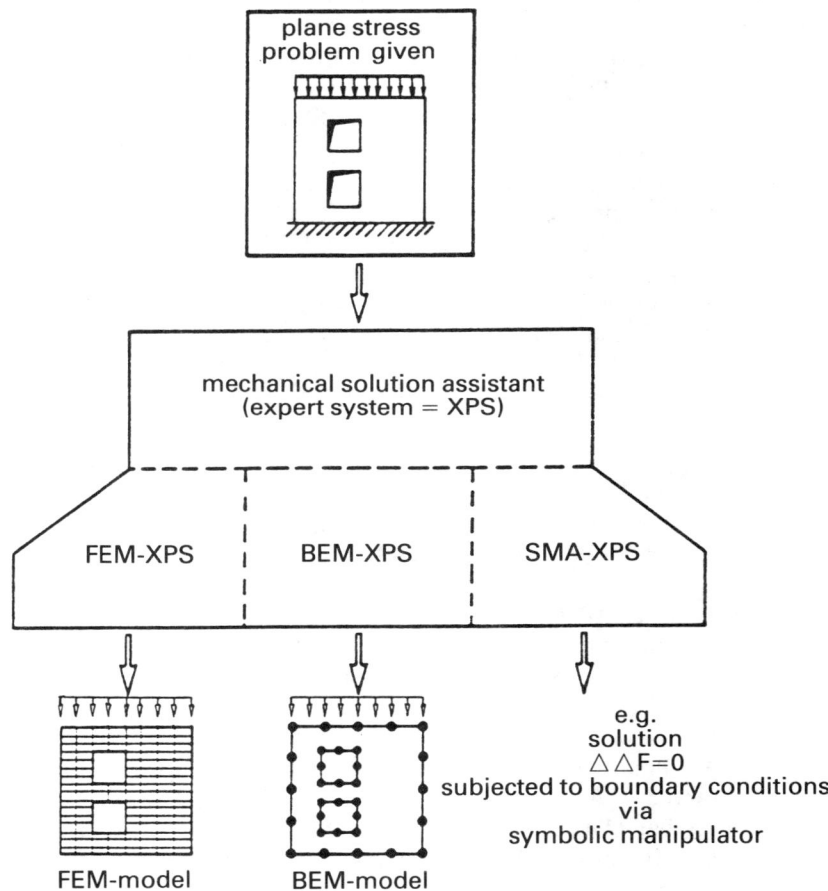

Fig. 6.1 — Main functionality of the mechanical solution assistant.

According to the given problem, the expert system queries facts from the engineer for classification (FEM, BEM or SMA). On the basis of the rules incorporated in the knowledge base, an appropriate solution to the problem is recommended, for example, a finite element approach associated with advice on the corresponding adequate finite element model including the complete input for a specified finite element program system.

The rules at the highest level of the knowledge base provide the information and facts needed for reasoning about which one of the main solution approaches is to be pursued (FEM, BEM or SMA). The following INSIGHT rules show an extract of top level knowledge:

RULE A1	classification level
IF	plane stress problem complicated
AND	represented by finite domains
THEN	finite element method (1.1)
AND DISPLAY	FEM example
RULE A2	classification level
IF	plane stress problem complicated
AND	represented by infinite domains
THEN	boundary element method (1.2)
AND DISPLAY	BEM example
RULE A3	classification level
IF	plane stress problem elementary
AND	symbolic manipulator available
THEN	symbolic manipulation approach (SMA)
AND DISPLAY	SMA example
EXPAND	plane stress problem elementary

Elementary plane stress problems are characterized by wall-like girders with non-disjunct domains such as
   simply supported rectangular slabs
   continuously supported slabs

Within the scope of this contribution, the finite element and boundary element approaches are not to be dealt with. By contrast, the symbolic manipulation branch is taken into consideration as this approach is also AI-related.

Symbolic manipulators constitute mathematical expert systems containing mathematical expertise about algebra and analysis. Simplification of expressions, differentiation, integration, solution of (differential) equations, matrix and tensor operations, etc., are performed by symbolic computations automatically rather than numerically. As a result, symbolic manipulation represents a new and powerful approach to applied mathematical problems (mechanical problems).

Here, it is to be assumed that the mathematical expert system (symbolic manipulator) operates separately from the mechanical assistant. Thus distributed expert systems are applied, each of which covers its own specific knowledge domain (see Fig. 6.2). At present, direct communication between the two expert systems is not manageable because of hardware and software limitations (probably realizable by fifth-generation computers). As

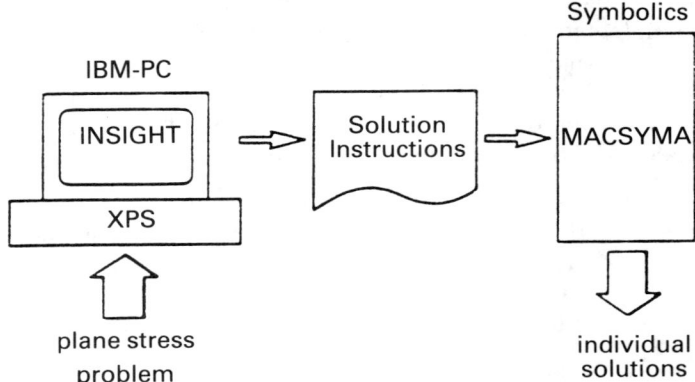

Fig. 6.2 — Distributed knowledge systems: INSIGHT–MACSYMA.

a consequence, the link between the two 'automatic experts' has to be accomplished by a human being (engineer).

As a result of the above-mentioned limitations, the knowledge base of the mechanical solution assistant only contains the basic instructions (and commands) to assist a user in handling the symbolic manipulator for plane stress problems. Therefore no manipulations are activated directly. Again, it should be emphasized that very complicated plane stress problems are preferably to be solved by finite computational methods (FEM or BEM). Closed-form solutions obtained by means of symbolic manipulators are currently still restricted to more elementary cases. Nevertheless, enhanced and improved manipulators may very soon be capable of solving even complicated partial differential equations associated with mixed boundary conditions. If that occurs, SMA will become a serious competitor to finite computational approaches.

To provide an insight into that part of the knowledge base that is navigated by the symbolic manipulator a simple plate subjected to in-plane forces (slab) is used to illustrate the capability of the assistant. The mechanical problem considered is shown in Fig. 6.3.

As demonstrated, the prescribed in-plane forces are determined by a cosine function. In particular, a restrain-free support is assumed. A solution of the plane stress problem must satisfy the equations of equilibrium and compatibility, and the boundary conditions. Combination of these yields the governing partial differential equation (body forces neglected)

$$\Delta\Delta F = 0 \qquad (1)$$

where F is the Airy stress function and Δ the Laplacian operator.

If the Airy stress function is known that also satisfies the boundary conditions, the unknown stresses

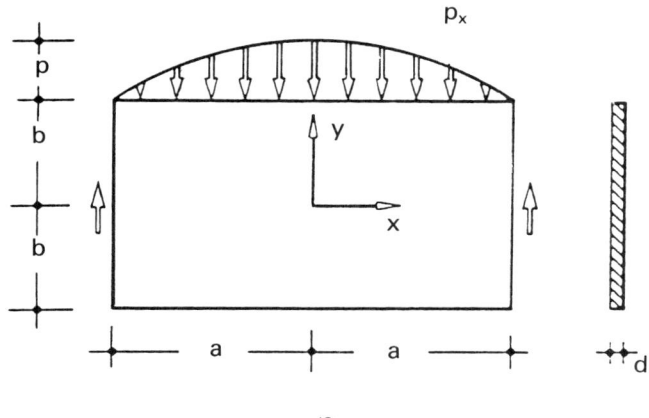

$p_x = p \cos \alpha x;\ \alpha = \pi/2a$

Fig. 6.3 — Elementary plane stress problem.

$$\sigma_{xx} = \partial^2 F/\partial y^2 \quad \sigma_{yy} = \partial^2 F/\partial x^2 \quad \tau_{xy} = \partial^2 F/\partial x\, \partial y \tag{2}$$

can be calculated. In our example, the boundary conditions take the form

$$\sigma_{xx}(x = \pm a, y) = 0 \tag{3}$$

$$\sigma_{yy}(x, y = +b) = (p_x/d)\cos(\alpha x) \tag{4}$$

$$\sigma_{yy}(x, y = -b) = 0 \tag{5}$$

$$\tau_{xy}(x, y = +b) = 0 \tag{6}$$

$$\tau_{xy}(x, y = -b) = 0 \tag{7}$$

Because of the cosine-distributed loading the boundary condition (3) can easily be satisfied. Hence, basically four remaining boundary conditions have to be taken into account.

To streamline the presentation, it is assumed that the appropriate Airy stress function is known. In this case the knowledge representation can be simplified; otherwise, a more comprehensive body of rules is required to identify the Airy stress function.

In our demonstrative example we have the function

$$F = \frac{1}{\alpha^2}[A \cosh(\alpha y) + B\alpha y \sinh(\alpha y) + C \sinh(\alpha y) + D\alpha y \cosh(\alpha y)]\cos(\alpha x) \tag{8}$$

This function is referred to as 'Airy stress function #1', and it contains four unknowns (A, B, C, D).

Then, the following INSIGHT production rule can be created to describe the symbolic manipulations needed to obtain the desired stresses according to eqn (2):

Rule	rectangular plate subjected to in-plane forces
IF	Airy stress function #1
THEN	symbolic manipulator instructions #1
AND DISPLAY	CONCLUSION #1

where 'DISPLAY CONCLUSION' gives the following instructions on the PC screen:

(1) Rectangular plate subjected to in-plane case; Airy stress function with four unknowns A, B, C, D

$$F = \frac{1}{\alpha^2}[A\cosh(\alpha y) + B\alpha y \sinh(\alpha y) + C\sinh(\alpha y) + D\alpha y \cosh(\alpha y)]\cos(\alpha x)$$

where $\alpha = \pi/2a$.

(2) Make use of symbolic manipulator to obtain the following partial derivatives:

$$\sigma_{xx} = \partial^2 F/\partial y^2 \qquad \sigma_{yy} = \partial^2 F/\partial x^2 \qquad \tau_{xy} = \partial^2 F/\partial x\, \partial y$$

(3) Simplify obtained expressions according to step (2).
(4) Satisfy boundary conditions by symbolic computing:

$$\sigma_{yy}(x, y = +b) = (p_x/d)\cos(\alpha x)$$

$$\sigma_{yy}(x, y = -b) = 0$$

$$\tau_{xy}(x, y = +b) = 0$$

$$\tau_{xy}(x, y = -b) = 0$$

where

$$p_x = p\cos(\alpha x) \qquad \alpha = \pi/2a$$

(5) Solve equation system in the unknowns A, B, C, D according to step (4) symbolically.
(6) Simplify obtained expressions for A, B, C, D.
(7) Compute $\sigma_{xx}, \sigma_{yy}, \tau_{xy}$ symbolically by substituting A, B, C, D into stress expressions according to step (2).
(8) Specify values for geometrical and loading parameters a, b, and p.
(9) Plot two-dimensional distribution of stresses at specific sections.

Similar application examples can be processed analogously. Therefore the knowledge about a multitude of individual problems can be permanently conserved despite the diversity associated with potential problems. Thus, in its final stage, the mechanical solution assistant represents a knowledge reservoir on plane stress problems that can be applied to practical problems, but also training and tutoring of human experts is facilitated. However, full benefit from the assistant is only gained if a symbolic manipulation facility is available. The symbolic manipulator applied is discussed in the next section.

It should be emphasized that the INSIGHT expert system consists of a knowledge base compiler that compiles the source knowledge base. The compiler translates the source material so that the run version runs faster than the non-compiled version.

5. SYMBOLIC MANIPULATION BY MEANS OF MACSYMA

As outlined in the previous section, the symbolic manipulation acts as an auxiliary expert system to the mechanical solution assistant. Analogously to pre- and postprocessors in finite element programs it may be interpreted as a secondary processor running in parallel to (but separately from) the INSIGHT assistant system.

At present, the MACSYMA symbolic manipulator is applied. MACSYMA is one of the most sophisticated symbolic manipulators available. Originally, it was developed for specific AI machines (Symbolics Computer Systems). Subsequently, MACSYMA derivatives have also been produced to run on the most popular engineering workstations, such as SUN and APOLLO. However, 15 megabytes of disk space and 4 megabytes of memory are required, with 8 megabytes of memory recommended. The mathematics expert MACSYMA applies automated mathematical expertise to problems requiring specialized solution techniques. MACSYMA carries out computations symbolically on the basis of the LISP program language. As a typical AI language LISP handles the lists that are ideally suited to describing manipulations in algebra and analysis. Lists in LISP represent mathematical terms and symbols. Equally, lists constitute a versatile means for the representation of mathematical rules, operators and operations.

Just as computers revolutionized numerical analysis, symbolic manipulation (e.g. based on LISP) will revolutionize automation of mathematical operations by performing symbolic computations. The most significant advantage compared with conventional numerical methods is that a complete insight into the nature of solutions is generated. By way of example, it is well known that finite element computations only provide a 'picture' of a specified parameter set. On the contrary, symbolic manipulation yields a closed-form solution that may be subsequently evaluated for multiple parameter sets.

In addition, MACSYMA provides capabilities to transfer symbolic solutions directly into FORTRAN code for numerical computing. Such a transfer is advisable because the running time of MACSYMA is not as good as that of FORTRAN. In fact, MACSYMA qualifies as a powerful preprocessor in numerical analysis, particularly in those cases where manual computation of mathematical expressions is cumbersome and error prone. A typical example is the computing of partial derivatives of complicated functions needed for the Jacobian matrix within Newton's method to solve non-linear equations.

For all the reasons indicated above, MACSYMA is used as an auxiliary processor to the mechanical solution assistant (Fig. 6.2). While the solution assistant primarily selects the appropriate solution philosophy associated with a precise definition of solution steps, MACSYMA provides the mathematical expertise to obtain the explicit solution of a mechanical problem. In other words, the solution assistant represents conceptual knowledge while MACSYMA captures mathematical knowledge.

To exemplify the MACSYMA capabilities the plane stress problem discussed above is again examined. In detail, it is demonstrated how the partial derivatives of the Airy stress function (see step (2) in 'DISPLAY conclusion') are symbolically computed. Then, the MACSYMA commands for solving the linear equation system of four equations in the four unknowns A, B, C, D are presented.

The **C2** statement represents the definition of the Airy stress function:

(C2) F:(A*COSH(ALPHA*Y)*COS(ALPHA*X)+
    B*ALPHA*Y*SINH(ALPHA*Y)*COS(ALPHA*Y)
    +C*SINH(ALPHA*Y)*COS(ALPHA*X)+
    D*ALPHA*Y*COSH(ALPHA*Y)*COS(ALPHA*X))/ALPHA^2

In line **C3**, the partial derivate for the stress σ_{xx} is specified

(C3) SIGMAXX:DIFF(F,Y,2),RATSIMP;

where **RATSIMP** causes a simplification of all arithmetic expressions and subexpressions created by symbolic differentiation. In line **C4**, a potential factorization is invoked:

(C4) FACTOR(%);

In lines **C2**, **C3** and **C4** the stress σ_{xx} is automatically computed in a closed form. The result is indicated in line **D4**:

(D4) COS(ALPHA X) (ALPHA 8 Y SINH(ALPHA Y) + 2 D SINH(ALPHA Y) +
    C SINH(ALPHA Y)
    + ALPHA D Y COSH(ALPHA Y) + 2 B COSH(ALPHA Y) + A COSH(ALPHA Y))

Analogously, the stresses σ_{yy} and τ_{xy} are obtained (lines **C5**, **C6**, **D6** and **C7**, **C8**, **D8** respectively:

(C5) SIGMAYY:DIFF(F,X,2),RATSIMP;

(C6) FACTOR(%);

(D6) - COS(ALPHA X) (ALPHA B Y SINH(ALPHA Y) + C SINH(ALPHA Y)
    + ALPHA D Y COSH(ALPHA Y) + A COSH(ALPHA Y))

(C7) TAUX:DIFF(F,X,1,Y,1),RATSIMP;

(C8) FACTOR(%);

(D8) - SIN(ALPHA X) (ALPHA D Y SINH(ALPHA Y) + B SINH(ALPHA Y) +
    A SINH(ALPHA Y) + ALPHA B Y COSH(ALPHA Y) + D COSH(ALPHA Y) + C
    COSH(ALPHA Y))

In line **C9**, the coordinate y is set to b (in MACSYMA notation, bb is used because the capital letter **B** and the lower-case letter **b** would be identical to each other):

(C9) Y:BB;

The **C9** manipulation is necessary to evaluate the first boundary condition $\sigma_{xx}(x, y = +b)$ that is specified in line **C10**:

(C10) G1:-P*COS(ALPHA*X)/DD = SIGMAYY,RATSIMP;

where **dd** corresponds to d according to the MACSYMA grammar. Line **D10** gives the result:

(D10) $-\dfrac{\text{P COS(ALPHA X)}}{\text{DD}}$ =
(- ALPHA BB COSH(ALPHA BB) D - SINH(ALPHA BB) C
- ALPHA B BB SINH(ALPHA BB) - A COSH(ALPHA BB)) COS(ALPHA X)

Line **C11** defines the boundary condition $\tau_{xy}(x, y = b)$. The symbolic evaluation is given in line **D11**:

(C11) G3:0=TAUXY,RATSIMP;

(D11) 0 = ((- ALPHA BB SINH(ALPHA BB) - COSH(ALPHA BB)) D - COSH(ALPHA BB) C
    + (-B -A) SINH(ALPHA BB) - ALPHA B BB COSH(ALPHA BB)) SIN (ALPHA X)

(C12) Y:-BB;

After setting $y = -b$ in line **C12**, the remaining boundary conditions $\sigma_{yy}(x, y = -b)$ and $\tau_{xy}(x, y = -b)$ are computed in lines **C13**, **D13** and **C14**, **D14** respectively:

(C13) G2:0=SIGMAYY,RATSIMP;

(D13) 0 = (ALPHA BB COSH(ALPHA BB) D + SINH(ALPHA BB) C
    - ALPHA B BB SINH(ALPHA BB) - A COSH(ALPHA BB)) COS(ALPHA X)

(C14) G4:0=TAUXY,RATSIMP;

(D14) 0 = ((- ALPHA BB SINH(ALPHA BB) - COSH(ALPHA BB)) D - COSH(ALPHA BB) C
    + (B + A) SINH(ALPHA BB) + ALPHA B BB COSH(ALPHA BB)) SIN(ALPHA X)

Thus, in line **C15**, the MACSYMA command for the symbolic solution of the equation system in the four unknowns A, B, C, D is presented:

(C15) LOESUNG:SOLVE([G1,G2,G3,G4],[A,B,C,D]),RATSIMP;

The solution is given in **D15**:

(D15) [[A =
$\dfrac{\text{(SINH(ALPHA BB) + ALPHA BB COSH(ALPHA BB))P}}{\text{(2 ALPLHA BB SINH}^2\text{(ALPHA BB)- 2 COSH(ALPHA BB) SINH(ALPHA BB) -2 ALPHA BB COSH}^2\text{(ALPHA BB)) DD}}$

$$B = \frac{(\text{SINH}(\text{ALPHA BB})) P}{(2\,\text{ALPHA BB}\,\text{SINH}^2(\text{ALPHA BB}) - 2\,\text{COSH}(\text{ALPHA BB})\,\text{SINH}(\text{ALPHA BB}) - 2\,\text{ALPHA BB}\,\text{COSH}^2(\text{ALPHA BB}))\,DD}$$

$$C = \frac{(\text{ALPHA BB}\,\text{SINH}(\text{ALPHA BB}) + \text{COSH}(\text{ALPHA BB}))\,P}{(2\,\text{ALPHA BB}\,\text{SINH}^2(\text{ALPHA BB}) - 2\,\text{COSH}(\text{ALPHA BB})\,\text{SINH}(\text{ALPHA BB}) - 2\,\text{ALPHA BB}\,\text{COSH}^2(\text{ALPHA BB}))\,DD}$$

$$D = \frac{\text{COSH}(\text{ALPHA BB}))\,P}{(2\,\text{ALPHA BB}\,\text{SINH}^2(\text{ALPHA BB}) - 2\,\text{COSH}(\text{ALPHA BB})\,\text{SINH}(\text{ALPHA BB}) - 2\,\text{ALPHA BB}\,\text{COSH}^2(\text{ALPHA BB}))\,DD} \quad]]$$

The elapsed time for the solution is about 30 s.

Further evaluations based on MACSYMA allow a complete solution of the given plane stress problem including a two-dimensional plot of the stress distributions. However, because of limited space, further operational steps carried out by MACSYMA are omitted here.

ACKNOWLEDGEMENT

The author would like to express his thanks to F. Wolfermann, Symbolics Computer Systems Inc., Düsseldorf, West Germany, for his assistance in providing all necessary MACSYMA support.

REFERENCES

[1] INSIGHT, *Insight Knowledge System Manual*, Level Five Research Inc., Melbourne Beach, FL, 1987.
[2] Hartmann, D., Selection and evaluation of structural optimization strategies by means of expert systems. In Niku-Lari, A. (ed.), *Structural Analysis Systems*, Vol. 5, *Expert Systems in Structural Analysis*, Pergamon, Oxford, 1986.

PART II
Design Optimization

7

On a knowledge-based user interface for the structural optimization system LAGRANGE

K. Schittkowski
Universität Bayreuth
R. Zotemantel
Messerschmidt–Bölkow–Blohm GmbH

1. THE STRUCTURAL OPTIMIZATION MODEL

For many reasons it is advantageous to combine the finite element technique with mathematical optimization algorithms. The finite element model describes the actual structure that is needed to calculate displacements, stresses, and eigenvalues. However, in most cases, it is necessary to obtain structures with certain properties such as the minimum weight, allowable stresses, frequency bounds. These properties depend on the cross-sectional areas and wall thicknesses of the elements, and are called **structural variables**. The answer for the structure depends on the values and distribution of these variables. To obtain such results — minimum weight for a feasible structure — we can use the trial-and-error method. That means we start with a given structure, then we change the variables by hand on the basis of our experience, and then carry out again an analysis. This method may lead to a satisfactory result after a number of cycles, but we do not know exactly whether the achieved design is the optimal design.

In this situation, mathematical optimization will help us to find optimal and feasible structures. The algorithms search systematically for the optimum on the basis of mathematical considerations. The engineer can influence this process by defining objective function, constraints, design variables, and fixed elements in a suitable way, and by selecting an efficient and robust algorithm. In general, the algorithm finds the desired result in few iterations. Unfortunately, no unique algorithm is known which solves all problems in structural analysis. For this reason, the programming system LAGRANGE offers different optimization strategies (seven in the current version). To choose the right algorithm, the engineer must have the knowledge and experience of the model, finite elements, and optimization calculations.

The standard form of the mathematical programming model is to minimize an objective function $f(x)$ with respect to the design variable x, where x is an element of the n-dimensional Euclidean vector space. The feasible region is designed by m inequality constraints of the form

$$g(x) \geq 0$$

In most cases, $g(x)$ is a non-linear function of the design variables which may vary between lower and upper bounds:

$$x_l \leq x \leq x_u$$

The aim of the optimization is to minimize (or maximize) the objective function. There are many possibilites for defining desirable objectives, e.g. the weight, heat transfer and costs. In LAGRANGE the minimum weight objective function is implemented, but it is easy to formulate any other function. In general, it is also possible to optimize structures with several objective functions leading to multicriteria problems which must be solved by methods of vector optimization.

The design variables may change during the optimization process and are divided into several groups:

— elements of the structure, i.e.
    cross-sectional areas of rods and beams,
    wall thicknesses of membrane and plate elements,
    laminated thicknesses of every layer in composite elements;
— balance masses;
— nodal coordinates.

While in sizing problems the shape of the structure does not change, the variations of grid points leads to a new structure within some given bounds. The program is able to optimize all mentioned design variables simultaneously. The sizing problem can be solved in the usual way as by any other structural optimization system, but special-purpose techniques were developed to realize shape optimization. To reduce the number of design variables, it is possible to fix some variables or to link variables. Fixing means that the design variables remain unchanged at their initial values. The linking option allows several design variables to be combined and treated as one design variable.

Constraints serve to guarantee a reasonable and realistic model. Manufacturing and physical conditions are natural bounds which restrict the choice of a feasible design. For all materials, the allowable stresses must not be violated and in the case of composite materials the strains have to be bounded (also called material constraints). Here manufacturing constraints mean that for composite elements one layer is bounded by a percentage of the whole thickness of the element.

By using gauge constraints in the form of lower and upper bounds, the design variables are forced to take meaningful values. The constraints restrict the design space and must be fulfilled simultaneously. If the design variables satisfy all constraints, the structure is called a feasible structure. Depending on the physical problem, many type of constraints and combinations of constraints are possible:

— the lower and upper bounds of the design variables (gauges);
— restrictions on displacements of nodal points;
— stress and strain constraints;
— aeroelastic efficiencies;
— flutter speed;
— constrained eigenfrequencies, eigenmodes;
— dynamic response;
— constraints concerning buckling.

Although, theoretically, the searched optimal solution must satisfy all constraints exactly, it is allowed in realistic structures to relax this requirement and to accept reasonable tolerances.

If vibrations are expected, it is necessary that the eigenfrequency is significantly lower or higher than other frequencies to avoid dangerous resonances. Sometimes it is useful to control a definite eigenmode which may change in the order of the eigenvalues. The aeroelastic efficiency is defined as the ratio of the aerodynamic forces working on the deformed structure to the aerodynamic forces on the undeformed structure. This value lies between 0 and 1. The aerodynamic forces change as a result of deformation of the structure and therefore the lift changes. The constraint is a given value of this efficiency. The flutter speed must be significantly higher than the maximum speed of the aircraft. Normally, the vibrations are damped, but they change with increasing speeds. At a certain speed — the flutter speed — the damping terms are overwhelmed by others and then the vibrational instability destroys the structure. Fluttering is a eigenvalue problem. Some structural elements such as plates and beams may tend to buckle. Therefore the stiffnesses have to guarantee that the stresses do not exceed the critical stress and that no instability is attained.

2. THE SYSTEM ARCHITECTURE

The programming system LAGRANGE is embedded in a CAD/CAE environment with several interactions with pre- and postprocessing features. It is designed in a modular way with predefined interfaces. The architecture is shown in Fig. 7.1.

We observe a strict separation between the input/output part and the optimization–analysis part. The link is represented by the IO subsystem and the control file. The IO subsystem consists of a database which realizes efficient data transfer between different LAGRANGE modules. The control file contains all important parameters such as number of nodes, elements and type of optimization strategy and enables the system to create

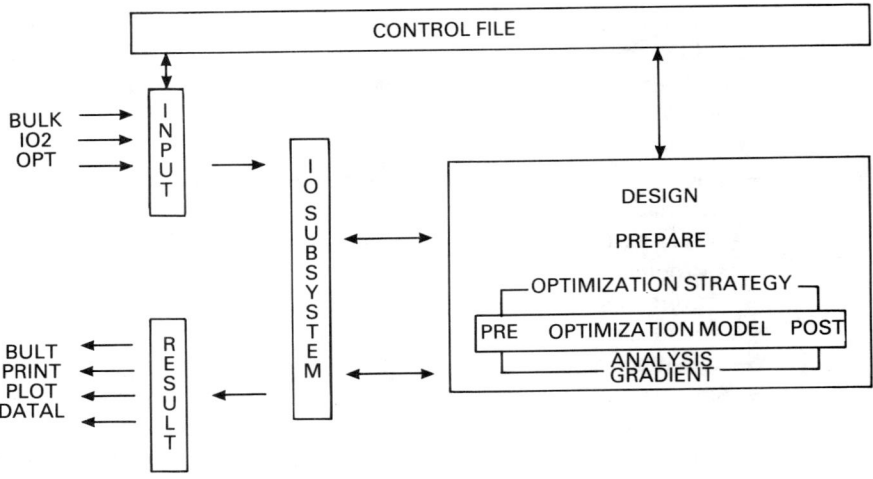

Fig. 7.1 — The architecture of the LAGRANGE system.

an excutable load module whose size is precisely adjusted to the given problem. The modules INPUT and RESULT are used to enter data or to prepare the optimization results for a suitable output, while the numerical optimization calculations are performed in the module DESIGN. The input data describing the finite element model can be provided in the form of a NASTRAN bulk data deck or an IO2 file created by NASTRAN. Supplementary input data describe the optimization model. In DESIGN we have separated optimization and analysis parts. The latter part contains, for example, the structural response and gradient calculation.

There are many possibilities for presenting the results and for using CAE systems available in LAGRANGE. As well as the standard output lists, it is possible to produce a bulk data deck for further NASTRAN applications and a data loader for SUPERTAB (IDEAS) of PATRAN to plot the optimal structure with many additional features, and to plot data for the documentation of the optimization history.

3. THE USER INTERFACE

The main menu of the user interface allows any of the subsequently explained system functions to be activated. By further submenus and questions, the user is guided through the system, so that the learning of a 'language' is not required. Whenever he is not aware of the set of allowed answers, he may type a '?' to obtain more detailed information. In most cases, it is possible to return immediately to the main menu by typing '<', e.g. when a severe input error was made or when the input is to be interrupted for some other reason.

3.1 New design

The command is used to generate a new optimization problem and to include it in the actual database. First, some information on the design to be optimized must be typed, e.g. the name which is used to identify the design in the database, some information for documenting the practical situation, and the name of the data file containing the geometrical problem description in a suitable format. After some consistency checks, a window displays the options that can be chosen to specify certain system facilities, e.g. minimization of bandwidth of output size. A submenu is then displayed for the constraint type to be specified. The corresponding numerical data which are required to define them, e.g. by upper and lower bounds, are inserted in form of a sequence of windows. Whenever possible, the given NASTRAN bulk data deck is investigated by the system, e.g. to search for certain cards and element numbers.

In addition, the design variables must be specified by defining the corresponding element and layer numbers either individually or in the form of ranges. Again, windows are displayed on the screen in tabular form to facilitate the input as much as possible. Simple consistency checks are performed at least to guarantee that the typed real or integer numbers are within their allowed ranges.

3.2 Display of design data

All design data and numerical results are kept in an internal database. On request they can be displayed on the screen individually. Moreover, lists of all available problems, together with some characteristic data, e.g. number of elements and design variables, and of the optimization history of a special design problem can be shown in form of tables. In the latter case, a user will be informed on all performed solution atttempts by indicating the chosen method, the used tolerances and parameters, the initial state, and the most important performance data, i.e. number of function and gradient evaluations and computing time. By offering the possibility of obtaining information on the whole solution process very conveniently, it is hoped that a user will update his own experience very rapidly as to which mathematical algorithm might be preferable in a given situation.

3.3 Output of the result file

When solving a structural optimization problem interactively, only very limited information is displayed on the screen to inform the user about the current state of the optimization process. Much more detailed information is stored on an output file, which can be investigated on the screen or sent to the line printer on request.

3.4 Edit design data

In principle, all data which are interpreted by a user to define a structural optimization problem may be edited on request, as may the data defining the finite element structure. For editing constraints or the variable linking of design variables, the user can select either the operating system editor which is executed automatically or the same windowing technique which was used before to define the problem data initially. In the first case, the formatted data are edited, and the user must be familiar with the feasible formats and commands. The corresponding section of the system documentation is, however, available and can be investigated on request. In the second case, known data may be edited directly in tabular form, data are added to existing data, or edited data replace the exisiting data depending on their structure and internal organization.

3.5 Start optimization run

LAGRANGE requires the generation of a specially formatted input file that contains some control information, the NASTRAN card deck, the definition of constraints and their corresponding data, the variable linking information, and some optimization parameters depending on the chosen solution method. By initiating the start command, this input file is generated automatically either for performing only an analysis for the initial design or to start an optimization run. Since many different optimization routines are available, a user must either define them 'by hand' or require a selection to be made by the system. In this case, a rule-based subprocess will send some questions to the terminal and, depending on the answers and the information on the design available so far, a heuristic proposal is made: cf. section 4 for details. A user may accept the proposed method and parameters or he may choose another code.

If some results obtained from a previous run with the same algorithm are available, it is possible to perform a warm start, i.e. continuation of the iteration which was interrupted before by exceeding the maximum number of iterations. Otherwise, a cold start may be activated starting from the last computed iterate or, alternatively, a new optimization cycle is initiated starting from the originally given design variables. Then a user determines the computer which is to be used to perform the numerical optimization. If necessary, the FORTRAN main program is generated automatically with precise dimensioning parameters, compiled, linked only with those modules needed to solve the problem, and executed immediately.

Alternatively, the user may prefer execution in the form of a batch job. In this case, solution data are stored on a system file and are sent to the database when LAGRANGE is started again. Otherwise some control information is displayed in the terminal in each interation, e.g. objective function and maximum constraint violation, and, after the optimization cycle has finished, the main menu can then be used foro initiating other system functions, e.g. the output of the achieved solution data which are available now in the database.

3.6 Failure analysis

LAGRANGE possesses a very flexible failure system and it is out of the scope of this report to explain all of its features. Severe failures interrupting the optimization are written to the output file mentioned before, and are sent to the terminal. By activating the failure analysis, a user will see the same failure information again. Subsequently, a rule-based, heuristic proposal of a suitable remedy is displayed and the user may accept the proposed action or not.

3.7 Delete a design

Any structural optimization problem generated by a user is kept in a database and can be deleted on request. Since all numerical data defining the geometry, the constraints and the variable linking are stored internally on files, a user will be asked whether he prefers to delete these files as well. Note that the user-provided geometry data in NASTRAN or IO2 format are still available in their original file.

3.8 Copy data from one optimization problem to another

In many cases, it might be desirable to generate a few different versions of one and the same structural optimization problem or to generate very similar design problems. In this case it is possible to copy data, e.g. for defining a special constraint type, from one problem to another. If the second problem is not available in the database, a new design problem is declared automatically where all design data are taken from the first. They can then be edited or modified in any way.

3.9 User documentation

The user documentation of the LAGRANGE system is part of the system itself. By choosing a suitable subcommand, a user may request the display of any section of the user documentation on the screen. Alternatively, the entire documentation can be printed on request on the available line printer.

3.10 Sort structural optimization problems

The principle key for identifying a design problem in the database is its name defined initially by the user. However, the sequence of all available problems in a database is not sorted at all. Thus a user might wish to sort them according to some information, e.g. name, date or user identification, to obtain a more readable list when starting the output command. The sorting key selected by a user does not influence the internal efficiency of the interface.

3.11 Direct input of operating system commands

When starting the user interface, a number of control procedures must be initiated, e.g. the internal linking of the data structure, the inclusion of known numerical data, and the preparation of the program to be interpreted. To avoid the necessary waiting time when a few operating system commands are to be executed within a session, it possible to insert them directly. Any operating system command may be typed then and the command is evaluated in the form of a separate subprocess.

3.12 Interactive access to database

The user interface of LAGRANGE is written in SUSY, an interpreter language specifically designed for the implementation of systems of the considered type; cf. [6]. Using some of the available SUSY commands, it is possible to act directly on the underlying database to perform actions that cannot be achieved by the predefined system functions. Since a user must be familiar at least partially with the structure of the database and the SUSY commands, he will be asked whether he wants to see the data structure on the screen or any section of the original SUSY documentation which is part of the system.

3.13 Halt

The halt function will save all data again on the database file that was declared when starting the system, and return the control to the operating system.

4. CHOICE OF AN OPTIMIZATION METHOD

The implementation of the LAGRANGE system is based on the idea of providing as many different optimization methods as available, and to have the possibility of exchanging, modifying, or extending them whenever it seems to be necessary on the basis of increasing practical experience. Therefore, the whole architecture of the system is highly modular and it is particularly easy to add a new optimization code. The reason for attempting to achieve a fairly large algorithm base is found in the observation that none of the codes tested so far in the frame of the LAGRANGE system is superior to all others with respect to efficiency or reliability. In addition, some codes are applicable only under certain conditions, e.g. feasible starting point or limited number of constraints and design variables.

The forecast, however, as to which of the available algorithms might be the most suitable in a given real situation is very difficult and can be made at most by experts, i.e. those engineers who did work with the system for a long time. Nevertheless, they may fail as well and the whole decision process is based on heuristics and some very vague criteria. However, since any false decision is, in most cases, cost expensive and delays the design process, there is a strong need to collect the experience of the experts and to implement it in a suitable way, so that their knowledge is available also for 'newcomers' who use the system for the first time.

In the present version of LAGRANGE, the following mathematical optimization algorithms are included and may be selected by a user:

IBF   Inverse barrier algorithm, i.e. a penalty-type method forming a sequence of unconstrained non-linear minimization problems. All iterates are feasible.

MOM   Method of multipliers, i.e. solution of a sequence of unconstrained non-linear optimization problems minimizing an augmented Lagrangian function.

SLP Sequential linear programming algorithm, i.e. solution of a sequence of linear programming problems obtained by linearizing the objective and constraint functions. The algorithm is stabilized by trust regions, certain bounds on the choice of a search direction that may change dynamically; cf. [3].

SQP1 Sequential quadratic programming algorithm, i.e. solution of a sequence of quadratic programming problems obtained by a quadratic approximation of the Lagrange function and a linear approximation of the constraints. The algorithm is stabilized by a line search with respect to an augmented Lagrangian merit function; cf. [5].

SQP2 Sequential quadratic programming algorithm, i.e. solution of a sequence of quadratic programming problems obtained by a quadratic approximation of the Lagrange function and a Linear approximation of the constraints. The algorithm is stabilized by a line search with respect to an exact penalty function and a 'watchdog' technique; cf. [4].

GRG Generalized reduced gradient method, i.e. elimination of dependent variables and projection of all iterates onto the feasible domain; cf. [1].

SRM Stress ratio method, i.e. heuristic approach based on optimality criteria and special constraints, e.g. stresses.

In addition, each algorithm requires the definition of certain tolerances and parameters that influence the solution process. There are practical situations in which any of the mentioned codes might be superior to the others. The underlying model structure is quite special, and favours the usage of the one or other code in a certain case. To give an example, consider the sequential linear programming method. This code should be used, for example, whenever

(i) there exist very many design variables and constraints;
(ii) many active constraints are expected to be active at the optimal solution;
(iii) stress constraints dominate.

However, the last two items are not definitely known in general, so that the conclusion that the SLP method should be used in this case depends on the degree of belief as to whether they are valid or not. Since the implementation of knowledge associated with uncertainties is hard to achieve in any conventional programming language, it was decided to implement it in form of rules and to use the inference mechanism of the SUSY language; cf. [6]. Thus one is able to formalize the known heuristics based on the numerical experience of all previous optimization runs, in form of rules. To give an example, a subset of those rules that favour the SLP method is listed here:

if NIL

```
          then slp-method with cfslp
!
!          Many active constraints
!
          if many-active-constr
          then slp-method with 50
!
!          Dominant stresses
!
          if dominant-stresses
          then slp-method with 30
!
!          Large optimization problem
!
          if large-design
          then slp-method with 30
!
!          Very large optimization problem
!
          if very-large-design
          then slp-method with 50
!
!          No static analysis required
!
          if design .. statanal <>/J/
          then not slp-method with 50
```

The values of the certainty factors range between 0 and 100, where 0 means that the action is not favoured, and 100 that the action is certain to be executed. When initiating the rule evaluation, the certainty factor of the action 'slp-method' obtains the value of the variable 'cfslp', which depends on the underlying self-learning process of the system and previous solution attempts of the design problem under consideration with the SLP method. Then this factor is updated according to those rules whose premises are satisfied in the actual situation. Either they are logical expressions depending on available data or they depend on other actions that describe a situation and that are obtained either from answers from the user of from the previous conclusions of other rules.

Performing the action 'slp-method' after evaluating all rules in the form of a loop means that a variable associated with the SLP method acquires the actual value of the accumulated certainty factor of the decision process. The corresponding values obtained for all included optimization codes are displayed to facilitate the decision of the user. He is of course allowed to reject the method that had the largest certainty value and to prefer another. For more details on the internal inference mechanism, the reader is referred to the SUSY documentation; see [6].

5. The SUSY language

The user interface of LAGRANGE is implemented in the SUSY language developed by Schittkowski [6], the numerical analysis part in FORTRAN. SUSY was designed to develop interactive software systems such as data management systems, interactive user interfaces, intelligent software systems (expert systems), or integrated problem solving systems. Since a SUSY program can interact directly with the operating system and therefore with existing programs written in any other language, a wide variety of possible applications exists.

The main important facilities of SUSY are as follows:

— permanent and temporary variables of various data types (CHAR, NAME, INTEGER, DECIMAL, STRING, LINE, TEXT, TABLE, FILE);
— structured data types (RECORD, STRUCTURE);
— assignment, compound and goto statements;
— logical and arithmetic expressions;
— brief input and output commands (':' — input from and output to terminal; '<' — output to file; '>' — input from file, '?' — display of help text);
— windows and input masks (WINDOW FROM ... TO ..., ENDWINDOW);
— database commands (NEW, LAST, FIRST, NEXT, DELETE, SEARCH, SORT, ...);
— table calculation (MIN, MAX, AVERAGE, SUM, ...);
— file management (RESET, REWRITE, SEARCH, SORT, ...);
— interactive input of SUSY commands (INTERACTIVE);
— include files and macros (INCLUDE, MACRO);
— reasoning processes (REASONING, RULES, ACTION);
— direct input and execution of arbitrary operating system commands ($ ⟨command⟩, GO, EXECUTE);
— help option (HELP, DOCUMENTATION);
— editor and programming environment (EDIT);
— text formatter (FORMAT);
— detailed error messages.

Large data sets can be processed internally by user-defined keys and index tables in the form of AVL trees. The reasoning process allows the storage and processing of heuristic knowledge based on certainty factors. A typical rule is of the form

$$\text{IF } \langle \text{antecedent} \rangle$$
$$\text{THEN } \langle \text{consequent} \rangle \text{ WITH } \langle \text{cf} \rangle$$

Antecedents can be logical expressions or actions, whereas the consequent must be an action, i.e. either a symbol defining a node in a decision tree or an arbitrary sequence of SUSY lines. The reasoning process allows additional bounds for the antecedent certainty factors, logical connections of antecedents, repeated execution of rules, automatic alteration of certainty factors in the execution of a rule, modification of certainty factors 'by hand', and an explanation component. The internal reasoning strategy is forward driven, executing the action that received the highest certainty factor within a loop. Backward chaining proceeding from goals is in preparation, together with alternative knowledge representations in the form of frames and object–attributable–value triples.

The SUSY interpreter is written in PASCAL and running in a VAX/VMS or UNIX programming environment. An interactive installation program facilitates the generation of a new software system written in SUSY.

REFERENCES

[1] Bremicker, M., *Entwicklung eines Optimierungsalgorithmmus der generalisierten reduzierten Gradienten, Andwendung auf Beispiele der Struckturdynamik*, Bericht, IMR, Univeristät Siegen, 1986.

[2] Kneppe, G., Krammer, J. and Winkler, E., Structural optimization of large scale problems using MBB–LAGRANGE, *Rep. MBB-S-PUB-305*, Messerschmidt–Bölkow–Blohm, Munich, 1987.

[3] Kneppe, G., *Direkte Lösungsstrategien zur Gestaltsoptimierung von Flächentragwerken*, VDI-Verlag, 1986.

[4] Powell, M. J. D., VMCWD: a FORTRAN subroutine for constrained optimization, *Rep. DAMTP 1983/NA4*, University of Cambridge, Cambridge, 1982.

[5] Schittkowski, K., NLPQL: a FORTRAN subroutine solving constrained nonlinear programming problems, *An. Oper. Res.*, **5**, 485–500, 1985/86.

[6] Schittkowski, K., *Die Systementwicklungssprache SUSY*, Bericht, Mathematisches Institut, Universität Bayreuth, Bayreuth, 1987.

8

Statistical machine learning for the cognitive selection of non-linear programming algorithms in engineering design optimization

D. A. Hoeltzel and **W. H. Chieng**
Columbia University

1. INTRODUCTION

Numerical optimization techniques, in the form of non-linear programming (NLP) algorithms, have been applied extensively to critical structural design and analysis problems for more than 30 years [1], and to a lesser entent to mechanical design problems [2].

The non-linear programming problem considered here takes the form

minimize:
$F(X)$
Subject to: ↑ objective function
$g_j(\mathbf{X}) \leq 0$ $j = 1,m$ inequality constraints
$h_k(\mathbf{X}) = 0$ $k = 1,l$ equality constraints
$X_1^l \leq X_i \leq X_i^u$ $i = 1,n$ side constraints

where $\mathbf{X} = \begin{bmatrix} X_1 \\ X_2 \\ \cdot \\ X_n \end{bmatrix}$ is a vector of design variables.

Numerical optimization provides a systematic, rational and directed approach to design decision making where, previously, heavy reliance was placed on the experience and intuition of the designer in achieving an improved design. Because of the complexities involved in the implementation of NLP algorithms, several researchers have undertaken performance analyses [3–5], the purpose being to determine correlations between the design problem type, the numerical optimization method and the corresponding results. On the basis of such studies, it is expected that the novice user should be able to understand better the capabilities of existing optimization methods and, furthermore, to utilize them without the need to undertake exhaustive programmes for testing and learning. While in concept this appears to be a rational approach to ascertaining the capabilities of a particular algorithm for a specific problem, in reality, as Himmelblau [6] states,

> a guarantee of convergence for an algorithm for special cases may offer little insight as regards satisfactory strategies for more complex problems

An optimization process invariably involves a trade-off between reality (completing and understanding the search process) and economy (evaluating a limited number of test functions). A process referred to as statistical concept learning (learning about new concepts by using given statistical measurements) is introduced to compensate for this trade-off. On the basis of well-organized data hierarchy, concept learning has been developed to eliminate unwanted knowledge which may result from noisy data (a small amount of data contradicting the conclusions which are agreed upon by a majority of the remaining data — in other words, data lying outside any of the defined cluster groups (Fig. 8.1) [7], and a scheme for the generalization of the statistical results has been developed.

2. METHOD SWITCHING STRATEGIES IN NON-LINEAR OPTIMIZATION

Existing algorithms for non-linear programming which have been surveyed [8,9] may converge to local optima which are not necessarily global optima. Many techniques for locating global optima, apart from knowing which method is the best first method, have yet to be uncovered. Method switching strategies are based, by analogy, on the game of golf (the reason for method switching is in accordance with the local geographical design information at the numerical optimum, and is analogous to the reason for selecting an appropriate golf club, in the game of golf, to strike the ball) rather than on the use of a one-step optimization scheme. This method switching procedure is designed to be one level higher than the so-called optimization strategy level [10] (monitors the numerical optimization process) and switches suitable numerical method combinations according to local design data.

For example, in the following problem containing a single objective function, design variable and constraint:

minimize the design objective function $\cos(x/1000 - 5.0)$
subject to the design constraint function $400.0 - x^2 \leq 0.0$
with design variable bounds $0.0 \leq x \leq 7500.0$

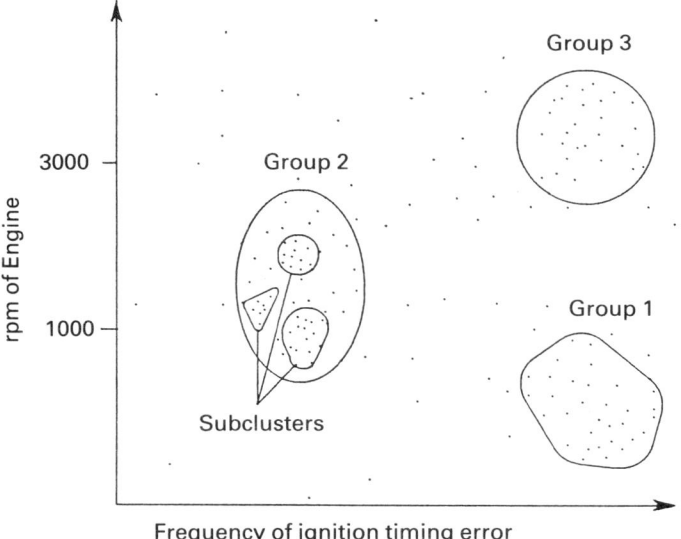

Fig. 8.1 — A clustering example: group 1 (low rotational frequency) and group 3 (high rotational frequency) cause more ignition failures; those points which are not enclosed within any of the groups constitute noisy data.

the following cases may possibly occur:
(1) When $|x| \geqslant 20$, the local information indicates that the design constraint is inactive.
(2) When $x \approx 5000$, the local information indicates that the objective function can be linearized to a polynomial of degree 2, which is $1-(x/1000-5)^2/2$.
(3) When $|x-5000|<10$, the local design information indicates that the objective function can be linearized to a polynomial of degree 4 by using an approximation of a Taylor series expansion.
(4) When $x \geqslant 7500$, one morfe design constraint is added from the design bounds, which can be expressed as $x \leqslant 7500$.

This example demonstrates that local design information can change in various ways when the updated state of the design variables (position) is altered. Method switching strategies are based on this phenomenon and may be likened to a monitoring or blackboard (a model in which all intermediate messages and results are displayed to the user and stored in a common area, called a blackboard) style decision making process. Method switching keeps track of the local optimization information and switches methods when the current method fails.

According to the schematic representation depicted in Fig. 8.2, the first design starting point, P1, lies in an infeasible design region and is far away from the globally optimal point. A temporary goal may be expressed as 'move the design into the feasible region as soon as possible' to increase the design efficiency. When the design 'converges' at a local optimum, P2, current NLP methods fail to move away from this point. In accordance with the local information found in the vicinity of P2, the method switching manager pins down another temporary goal which may be stated as 'find a feasible design with a smaller objective value'. Method switching terminates when the convergence criteria have been satisfied. This is usually based on (1) a CPU time consumption limitation, (2) the number of algorithm iterations, or (3) relative or absolute difference between successive values of the objective function.

3. SAMPLE PROBLEM TESTING

Fifteen different attributes have been chosen to characterize the test of a sample problem. The sample problems can be separated into three domains:
(1) The design problem type contains eight parameters, including the number of design variables, the number of total design constraints, the number of equality design constraints, the number of active inequality design constraints, the maximum (positive) order of testing polynomials, the minimum (negative) order of testing polynomials and the function evaluation cost for one design function evaluation.
(2) The choice of non-linear programming method contains three parameters, which according to the automated design synthesis (ADS) numerical optimization library [10] are strategy, optimizer and one-dimensional search method.
(3) The performance of the result contains four parameters, including the minimum objective value reachability, the design constraint violation condition and the maximum distance of search.

The set of test problems for the learning program has been produced by a random function generator (Fig. 8.3), which randomly selects a problem type, and in accordance with the selected problem type generates the objective function and the design constraint equations. The polynomials can be thought of as local information in real-world design problem formulations since many functions can be expressed in a Taylor series expansion. Non-linearity, discontinuity and differentiability can be altered by appropriately adjusting the order of the polynomials.

After implementing these concepts using the ADS numerical optimization library, design problems have been tested by a number of method combinations, which have been randomly selected. The authors have generated approximately 10 000 samples with results using an IBM PC/AT microcomputer. These results have been subsequently analysed, using statistical machine learning concepts incorporated within a program referred to as OPTDEX–OLDM (optimum design expert–optimization level design manager), on a Symbolics 3640 AI workstation.

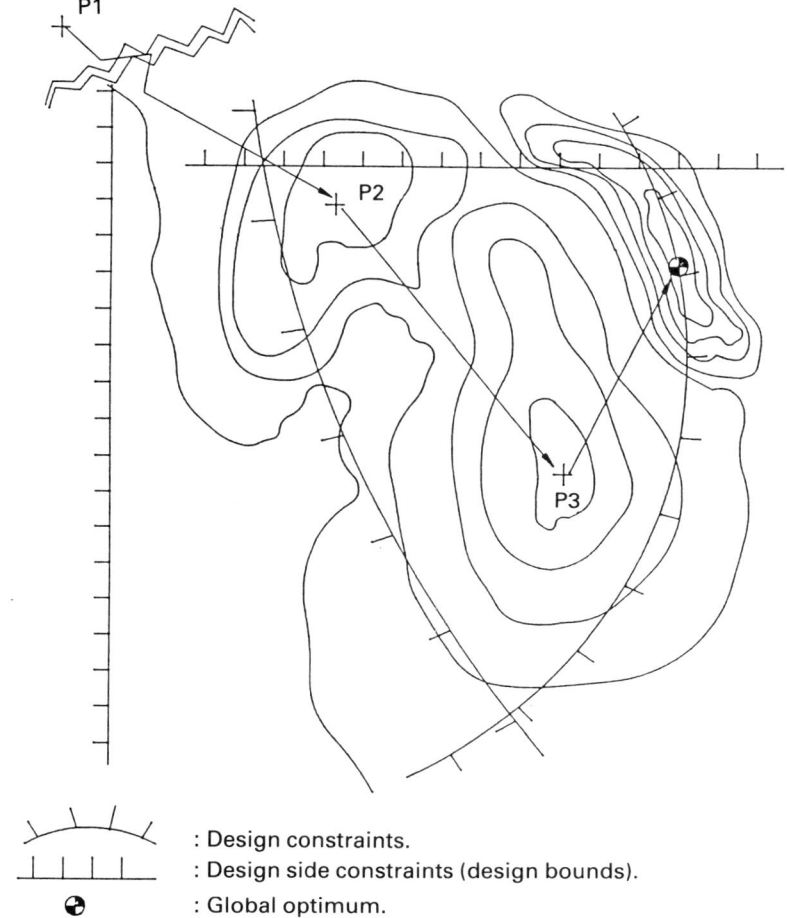

Fig. 8.2 — An example which demonstrates that local design information is different for different design starting points.

4. CLUSTERING AND ASSOCIATED STATISTICS

Every sample inherently has several attributes, which include the characteristics of the design problem type, the category of the non-linear programming method and the corresponding result. All of these attributes are represented quantitatively and some of them are noisy, i.e. unreliable. To minimize the noise factor, a 'variance' type of analysis [4] has been employed.

Clustering techniques [13–16] are used to find groups of samples, whose common characteristics have not been predefined. The aim is to subdivide the available samples into a relatively small number of groups, based on the statistical behaviour of the different attributes.

The clustering analysis involves the following concepts.

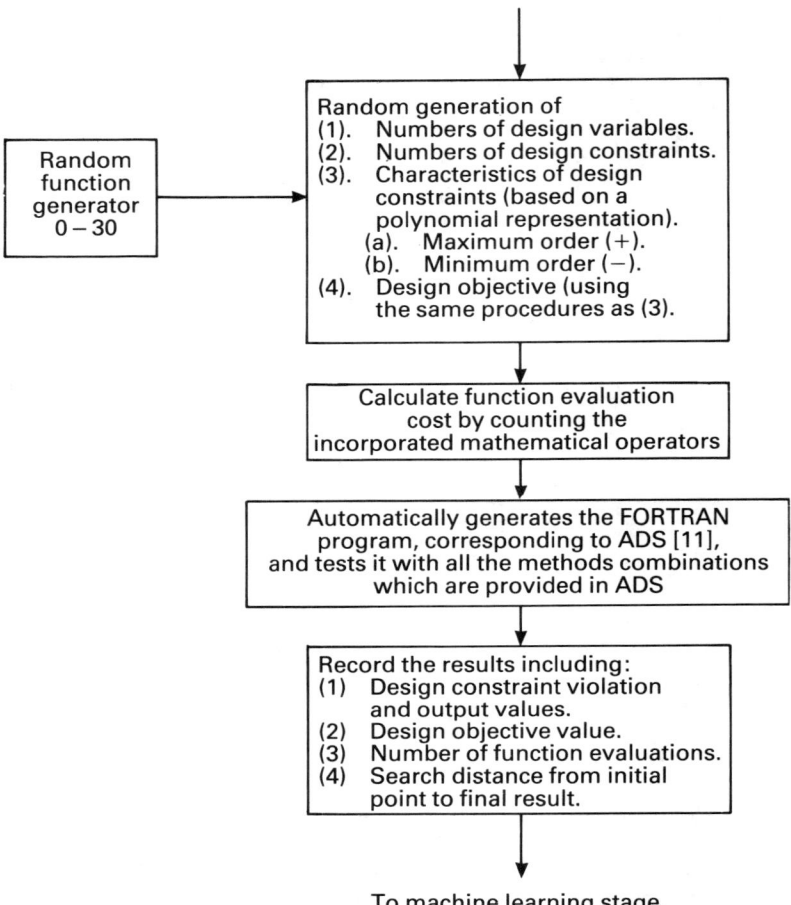

Fig. 8.3 — Flow control for random sample generation and testing.

4.1 Scaling

Scaling transforms the real-world value of each attribute into a machine-understandable scale. This can be done by calculating the mean

$$\mu = \frac{1}{m} \sum_{i=1}^{m} a_i \qquad (2a)$$

and the standard deviation

$$\sigma = \{E[(A - \mu)^2]\}^{1/2} \qquad (2b)$$

where m is the total number of data samples, a_1 the value of the ith attribute,

A the random variable which can assume the value a_i and E the expected value (statistical sense).

Various models may be chosen to represent the statistical distribution of the attributes. For example, if a Gaussian distribution is chosen, then 68% of samples will be distributed within one standard deviation about the mean, μ, and about 95% of the samples will be distributed within two standard deviations about the mean, μ. According to the mean, μ, and the standard deviation, σ, found for each variable, all the variables are normalized and digitized to a predefined scale. For the purposes of this research, a scale of 0–9 has been selected.

4.2 Non-hierarchical clustering

Non-hierarchical clustering is based on the optimization of a given grouping of objective functions, and represents the minimization of the sum of the variances within each group and the maximization of the sum of variances between groups:

$$\min_{C \in p(n,M)} \sum_{j=1, i \in C_j}^{n} \|a_i - a_j\|^2 \tag{3}$$

$$\max_{C \in p(n,M)} \sum_{j=1}^{n} m_j \|\bar{a}_j - \bar{a}\|^2$$

where $C = (C_1, C_2, C_3, \ldots, C_n)$ and C_i represents the ith cluster group, $M = \{1, 2, 3, \ldots, m\}$ is the set of all samples, $p(n,M)$ the set of all cluster groups C of M having length n, n number of cluster groups, $1 \leq n \leq m$, $\bar{a}$ the expected value of the total sample of attributes, $\bar{a}_j$ the expected value of C_j and m_j the number of samples in C_j.

Since the total scatter in a fixed sample size is constant [10], it is sufficient to minimize the sum of variances, $W(n, M)$, within each group. Therefore, eqn (3) can be expressed as follows:

$$\min W(n,M) = \min_{C \in p(n,M)} \sum_{j=1, i \in C_j}^{n} \|a_i - a_j\|^2 \tag{4}$$

The necessary tools for the clustering process are described below.
To calculate the new mean value from two given groups,

$$\bar{a}_{p+q} = \frac{1}{m_p + m_q}(m_p \bar{a}_p + m_q \bar{a}_q) \tag{5}$$

and to calculate the objective value (sum of variances) of the two given groups,

$$W_{p+q} = W_p + W_q + m_p[\bar{a}_p - \bar{a}_{p+q}][\bar{a}_p - \bar{a}_{p+q}]^T + m_q[\bar{a}_q - \bar{a}_{p+q}][\bar{a}_q - \bar{a}_{p+q}]^T \quad (6)$$

4.3 Clustering strategy

Since the number of all possible cluster group combinations (total clustering) can become prohibitively large, it is imperative that a reduction in the number of clusters be attempted. For example, suppose that m samples (attribute values) have to be clustered into n groups or less. This number of clusters is given by

$$H(n, m) = \frac{1}{n!} \sum_{i=1}^{n} (-1)^{n-i} \binom{n}{i} i^m \quad (7)$$

For $m = 1000$ and $n = 15$, $H(n,m)$ is greater than 1×10^{30}. For this research, $m = 10\,000$ and $150 \leq n \leq 300$, and therefore the clustering is not practically achievable. As a result, a special strategy has been employed to alleviate this problem. Instead of searching for total clustering, the OPTDEX–OLDM program starts from m samples and allows each single sample to be a group, i.e. $n = m$. The program then attempts to decrease the total number of cluster groups, during each clustering cycle, by one. During each cycle, the program searches for any two groups from the current set which satisfy the criterion of eqn (4). This clustering process terminates when the number of groups, denoted by n^*, satisfies the following condition:

$$\min W(n^*, M) \geq W_{\text{acceptable}} \quad (8)$$

Based on this clustering strategy, $H'(n,m)$, the reduced number of clusters, is

$$H'(n,m) \approx (m^2 - n^2)/2 \quad (9)$$

For the $m = 1000$ and $n = 15$ cases, $H'(n,m) \approx 5 \times 10^5$. Although noise may bias this type of clustering in the very early stages of processing, as previously predicted, when compared with the increased efficiency, by a factor of approximately 2×10^{12}, it is an acceptable strategy. Flow control for this process is shown in Fig. 8.4.

5. EXPLANATION OF THE STATISTICAL RESULTS

An explanation facility [17] is an important feature which distinguishes artificial intelligence programs from usual programs. Its purpose is to present the computational results in the form of a natural language that is

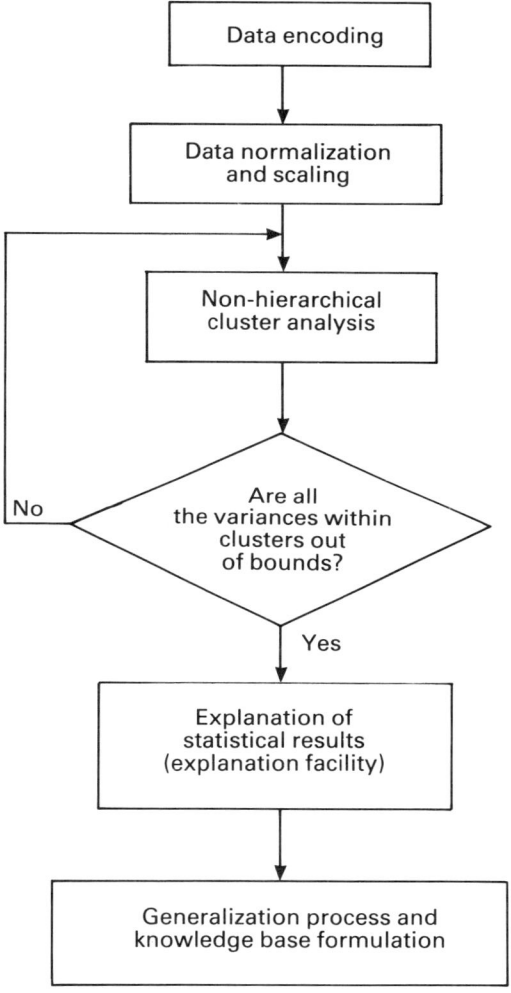

Fig. 8.4 — Flow control for cluster analysis.

comprehensible to a novice user. In addition, this capability forms the basis of incremental machine learning. A simple example that demonstrates how machine learning provides an explanation for a resulting cluster group follows.

Group 1.   Number of members = 17

Attribute	Range	Mean	Variance
Non-linearity	0–9	8	3.0
Strategy	0–9	2	0.2
Distance of Search	0–9	1	1.0

Response from the OLDM:

OLDM> I found that (as supported by 17 samples),
   If *Generally-speaking*, the nonlinearity is *very-high*, and
       Definitely, the strategy is the *linear extended interior penalty function method*
   Then *Most-likely*, optimization searching will be *very-local*.

(Italicized explanations represent terminology derived from the statistical results.)

6. CLASSIFICATION AND INCREMENTAL MACHINE LEARNING

Automatic concept learning, implemented in the form of concept learning generalization (the automatic generalization of a concept based on a sufficiently large number of agreements among specific case (non-general) concepts — in other words, expanding a concept to include a more general class of specific cases than previously included), has been shown to be useful in interpreting and organizing large amounts of information about a domain [12]. After performing the initial clustering from the test samples (tens of thousands of samples in this case), the OPTDEX–OLDM program reaches approximately 500 conclusions. These conclusions may overlap one another, some of them may be redundant and they all have to be appropriately formatted into a rule-based expert system.

Creating a classification scheme is typically the first step in developing the heuristics (rules of thumb) for a collection of observations or phenomena. The goal off the classification scheme is to structure given observations into a hierarchy of meaningful categories [6]. The OLDM applies generalization-based memory to build up a hierarchy of conclusions. It actually constructs a connective network to derive conclusions in a canonical form. A detailed explanation of this process is provided by Lebowitz [7]. An important feature of the OLDM is its ability to manage contradictions between conclusions, referred to as noise, by simply counting the number of supporting members for each conclusion. For example, the following conclusions (non-generalized) have been drawn by the OLDM:

Conclusion 1. *Supported by* 19 *members*
   If the Discontinuity is *high* and
       the Optimizer-choice is *Golden-section-method*
   then the objective value is *less-minimized*.

Conclusion 2. *Supported by* 25 *members*
   If the Discontinuity is *low* and
       the Optimizer-choice is *Golden-section-method*
   then the objective value is *less-minimized*.

Conclusion 3. *Supported by* 4 *members*
   If the Discontinuity is *high* and
       the Optimizer-choice is *Golden-section-method*
   then the objective value is *minimized*.

The generalized concept, drawn by the OLDM, based on these conclusions is

OLDM› CONCEPT-008:
        If the Discontinuity is *high or low* $^+$ and
  comment: $^+<$ this result is based on the generalization of conclusions 1 and 2 >
        the Optimizer-choice is *Golden-section-method*
        then the objective value is *less-minimized*. $^{++}$
  comment: $^{++}<$ the number of members supporting conclusion 1 is greater than the number supporting conclusion 3 >.

Another important feature of the OLDM is its ability to perform on-line statistically incremental machine learning. The OLDM is an on-line consultant during numerical optimization processing which has been incorporated within the ADS optimization library. According to the existing rules and local information from updated optimization searching, it chooses and switches method combinations from ADS and feeds back the result of each applied rule. These feedbacks are always represented in a standardized format with 14 parameters as previously described. Each piece of standardized information can be treated as an additional test sample, a_e, clustered into a group, C_j, which satisfies the following condition:

$$\min_{C \in p(n,M)} \frac{m_j}{m_j+1} [a_e - \bar{a}_j][a_e - \bar{a}_j]^T \tag{10}$$

During the incremental machine learning process, any of the existing cluster groups, say C_k, such that $W_k > W_{\text{acceptable}}$, has to be reclustered by utilizing the procedures which have been discussed. After the reclustering process has been completed, new concepts (conclusion) are born and/or old concepts die. This is referred to as the birth and death procedure for maintaining and renewing concepts in the knowledge base.

7. CONCLUSION

A new approach to design optimization, referred to as cognitive method switching, using non-linear programming algorithms applied sequentially, based on local design information, has been presented. Statistical evaluation with clustering of attributes associated with a randomly generated problem sample database, containing over 10 000 samples, has led to the generation of guidelines for the application of NLP algorithms to design optimization problems. Continued expansion of the problem database should permit more generalized guidelines to be obtained and thereby assist the non-expert user in cognitively selecting an appropriate sequence of NLP algorithms for a specific design optimization problem.

ACKNOWLEDGEMENTS

This research was supported by grants from the IBM and NCR corporations and by the US Army Research Office–DoD University Research Instrumentation Program under Grant DAAL03-86-6-0116.

REFERENCES

[1] Vanderplaats, G. N., *Numerical Optimization Techniques for Engineering Design with Applications*, McGraw-Hill, New York, 1984.
[2] Siddall, J. N., *Optimal Engineering Design: Principles and Applications*, Dekker, New York, 1982.
[3] Schittkowski, K., *Nonlinear Programming Codes, Lecture Notes in Economics and Mathematical Systems*, Springer, Berlin, 1980.
[4] Powell, M. J. D. (ed.), *Nonlinear Optimization 1981*, Academic Press, New York, 1982.
[5] Sandgren, K. M. and Ragsdell, P. The utility of nonlinear programming algorithms: a comparative study — parts I and II, *J. Mech. Des. ASME*, 540–551, 1980.
[6] Himmelblau, D., *Applied Nonlinear Programming*, McGraw-Hill, New York, 1972.
[7] Lebowitz, M., Concept learning in a rich input domain: generalization-based memory. In *Machine Learning: an Artificial Intelligence Approach*, Vol. II, Morgan Kaufmann, Los Altos, CA, 1986, pp. 193–214.
[8] Wilde, D., *Globally Optimal Design*, Wiley, New York, 1978.
[9] McCormick, G. P., *Nonlinear Programming: Theory, Algorithms, and Applications*, Wiley, New York, 1983.
[10] Vanderplaats, G. N., ADS—Fortran program for automated design synthesis, *NASA Contract Rep. 172460*, 1984.
[11] Spath, H., *Cluster Dissection and Analysis*, Ellis Horwood, Chichester, 1985.
[12] Cochran, W. G., *Sampling Techniques*, Wiley, New York, 1977.
[13] Green, W. R., *Computer-Aided Data Analysis — a Practical Guide*, Wiley, New York, 1985, pp. 185–202.
[14] Diday, E., Problem of clustering and recent advances. In J. Janssen (ed.), *New Trends in Data Analysis and Applications*, 1983, pp. 167–182.
[15] Mandel, J., *The Statistical Analysis of Experimental Data*, Dover Publications, New York, 1964.
[16] Stepp, R. E. and Michalski, R. S., Concept clustering inventing goal-oriented classifications of structured objects. In *Machine Learning: an Artificial Intelligence Approach*, Vol. II, Morgan Kaufmann, Los Altos, CA, 1986, pp. 193–214.
[17] Hoeltzel, D. A. and Chieng, W. H., An adaptive generic planning model for large scale integrated engineering design, *1st Eurographics Workshop on Intelligent Computer-Aided Design*, Springer, Berlin, 1987, Chapter 5.

9

A knowledge-based expert system for selection of slab type for multistory buildings

M. L. Das
University of Lowell
S. K. Ghosh
Portland Cement Association

1. INTRODUCTION

Design of the multistory buildings requires procedures based on both rules of thumb of heuristics and algorithms. Algorithms based on incontrovertible physical law alone are not adequate to produce an economical and realistic design.

During the last two decades, large numbers of reliable and sophisticated computer programs based on algorithms have been developed. Unfortunately, however, not enough work has been done to formalize the heuristics used in the industry to develope knowledge-based expert systems. There are two major problems which are responsible for this. First, the experts use heuristics from their experience in their domain of expertise without formalizing them as a system. Second, the experts are not committed to sharing their hard-earned experience with others.

This chapter attempts to emphasize the acquisition and formalization of heuristics (knowledge) and the creation of a knowledge-based expert system with these formalized heuristics.

At present, various inexpensive microcomputer-based expert system shells are commercially available. One of these available shells, EXSYS, is used to create this expert system and to provide dialogue with the users.

2. CAST-IN-PLACE FLOOR TYPES

Cast-in-place floor types which are generally used in multistory buildings are

(1) slab on beams (one way or two way)
(2) one-way joist
(3) flat plate
(4) waffle slab.

160 DESIGN OPTIMIZATION [Pt. II]

Each of these types has unique functions. Judicial selection of one of these types for a particular multistory building may result in substantial savings in construction cost. The present chapter is an attempt to help relatively inexperienced engineers to select a floor type ideal for their particular type of multistory building. Figs 9.1–9.5 illustrate these slab types.

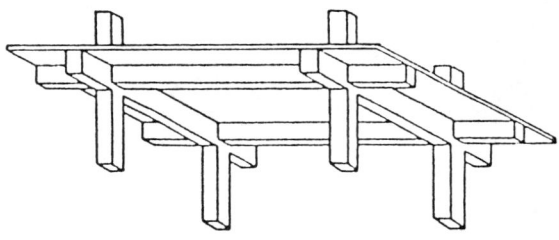

Fig. 9.1 — Slab on beams.

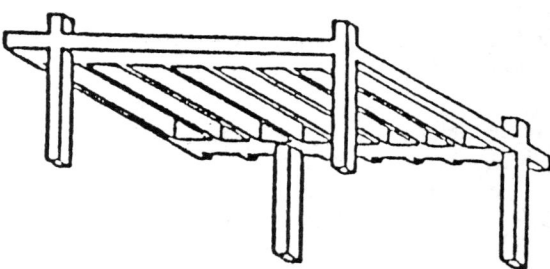

Fig. 9.2 — One-way joist.

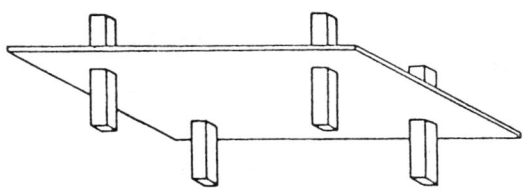

Fig. 9.3 — Flat plate.

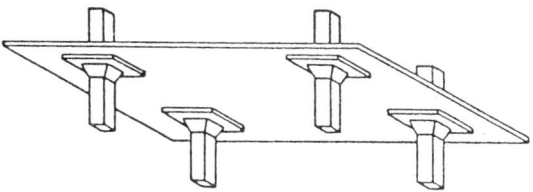

Fig. 9.4 — Flat slab.

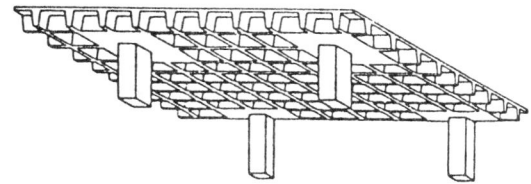

Fig. 9.5 — Waffle slab.

3. EXPERT SYSTEM SHELL

A microcomputer-based commercially available expert system shell, EXSYS, is used to create the present expert system. EXSYS is a rule-based system with IF, THEN and ELSE clauses. In this expert system only IF and THEN clauses are used.

The IF clause of EXSYS is made up of two parts, a qualifier and a value. The qualifier is usually the part of the condition up to and including the verb. The value is the possible completion of the sentence started by the qualifier. The THEN part is made up of choices with their corresponding importance factors. These importance factors are referred to as probabilities in EXSYS. To illustrate this concept, rule 3 of this expert system is quoted below:

IF:
    Lateral load carried by floor is wind & US seismic zone 1

THEN:
	slab on beams	— probability=21/100
and	one-way joist	— probability=23/100
and	flat plate	— probability= 9/100
and	flat slab	— probability=10/100
and	waffle slab	— probability=17/100

The IF clause in this rule is 'Lateral load carried by floor is wind & US seismic zone 1'. 'Lateral load carried by floor is' is the qualifier and 'wind &

US seismic zone 1' is the value. The THEN part is made up of five choices with their importance factors, referred to here as probability factors. These importance factors are based on a total possible score of 100.

EXSYS uses both forward and backward chaining algorithms to derive its conclusion. For this expert system only backward chaining is used. EXSYS provides three methods of calculating final combined importance factors. The first is a simple average of all of the individual importance factors of the choice received in the rules found to be true. The second combines the importance factors as if they were dependent probabilities. The third way to combine the factors is as if they were independent probabilities. The first method is selected for this expert system.

4. CHOICES USED IN THE EXPERT SYSTEM

There are five choices which are used in creating the rules of this expert system presented in this chapter:

(1) slab on beams,
(2) one-way joist,
(3) flat plate,
(4) flat slab
(5) waffle slab.

5. KNOWLEDGE ACQUISITION

The knowledge used in this expert system is from the long professional experience of both authors. This knowledge can be organized in the following functional–physical categories. The criteria specified in these categories are generally considered in the selection process of a floor system:

(1) lateral load resistance;
(2) ratio of adjacent spans in the same direction;
(3) panels are squarish in shape;
(4) longer panel dimension is reasonably short;
(5) live load is reasonably light;
(6) minimizing building height is important;
(7) a ready ceiling is important;
(8) simplicity of form work is important;
(9) flexibility of partition is important;
(10) easy poke-through hole for wiring is important;
(11) installation of vibratory machine.

5.1 Importance factors within categories

The eleven functional–physical categories just specified are then expanded for the various possible values a floor may assume within a particular category. Associated importance factors are also developed for each of the

SELECTION OF SLAB TYPE FOR MULTISTORY BUILDINGS

five choices. If a choice is an ideal, an importance factor of 100 is assigned for that choice. The following table is prepared on the basis of this concept.

		Items	Slab on beams	one-way joist	Flat plate	Flat slab	Waffle slab
(1)		Lateral load resistance					
	(a)	No lateral load	100	100	100	100	100
	(b)	Wind+seismic=0^a	90	100	40	50	75
	(c)	Wind+seismic=1	90	100	40	45	75
	(d)	Wind+seismic=2	90	100	35	35	60
	(e)	Wind+seismic=3	90	100	0	0	0
	(f)	Wind+seismic=4	90	100	0	0	0
(2)		Ratio of adjacent spans in the same direction					
	(a)	1.0	85	85	100	100	100
	(b)	1.33	100	100	100	100	100
	(c)	2.00	100	100	50	50	50
(3)		Panels are squarish in shape, with a long:short ratio of					
	(a)	1.0	80	60	100	100	100
	(b)	1.1	100	60	95	95	95
	(c)	1.2	100	70	85	85	85
	(d)	1.3	100	80	75	75	75
	(e)	1.4	100	90	65	65	65
	(f)	1.5	80	100	50	50	50
	(g)	1.6	80	100	40	40	40
	(h)	1.7	80	100	30	30	30
	(i)	1.8	80	100	20	20	20
	(j)	1.9	80	100	10	10	10
	(k)	2.0	80	100	0	0	0
(4)		Longer panel dimension is reasonably short					
	(a)	<13.5 ft (4.1 m)	60	50	100	30	20
	(b)	13.5–16.5 ft (4.1–5.0 m)	75	70	100	35	25
	(c)	16.5–19.0 ft (5.0–5.8 m)	85	90	100	50	35
	(d)	19.0–21.8 ft (5.8–6.6 m)	90	100	100	90	75
	(e)	21.8–24.5 ft (6.6–7.5 m)	90	100	80	100	95
	(f)	24.5–27.5 ft (7.5–8.4 m)	90	100	65	90	100
	(g)	>27.5 ft (8.4 m)	90	100	0	25	60
(5)		Live load is reasonably light					
	(a)	<50 lbf ft^{-2} (<2.39 kPa)	60	60	100	20	10
	(b)	50–60 lbf ft^{-2} (2.39–2.87 kPa)	70	70	100	25	15
	(c)	60–70 lbf ft^{-2} (2.87–3.35 kPa)	80	75	100	50	40
	(d)	70–80 lbf ft^{-2} (3.35–3.83 kPa)	90	80	90	100	80
	(e)	80–90 lbf ft^{-2} (3.83–4.31 kPa)	100	100	70	95	100
	(f)	90–100 lbf ft^{-2} (4.31–4.79 kPa)	90	100	60	85	100
	(g)	100–110 lbf ft^{-2} (4.79–5.27 kPa)	80	100	40	60	100
	(h)	110–120 lbf ft^{-2} (5.27–5.75 kPa)	70	100	30	40	100
	(i)	120–130 lbf ft^{-2} (5.57–6.22 kPa)	60	100	0	10	100

164   DESIGN OPTIMIZATION [Pt. II]

(j)		<130 lbf ft^{-2} (<6.22 kPa)	50	100	0	0	100
(6)		Minimizing building height is important					
	(a)	Local code limits height	40	50	100	80	50
	(b)	Saving cladding is important	40	50	100	80	50
	(c)	Saving power cable is important	40	50	100	80	50
	(d)	Saving telephone cable is important	40	50	100	80	50
	(e)	Saving HVAC^b is important	40	50	100	80	50
	(f)	Reducing heating and cooling cost is important	40	50	100	80	50
	(g)	Reducing window cleaning is important	40	50	100	80	50
(7)		A ready ceiling is important					
	(a)	Reducing hung ceiling cost is important	40	50	100	70	50
	(b)	Head room loss by hung ceiling is unacceptable	40	50	100	70	50
(8)		Simplicity of form work is important					
	(a)	Minimizing construction time is important	50	80	100	65	75
	(b)	Form work labour to material cost is high	50	80	100	65	75
	(c)	Use of flying form is important	20	75	100	75	40
(9)		Flexibility of partition is important	70	85	100	100	100
(10)		Easy poke-through hole for wiring is important	70	70	100	70	50
(11)		Installation of vibratory machine	100	80	30	40	80

^{a}Seismic=0–4 are the US seismic zones 0–4 as shown in the Seismic Risk Map of the United States.
^{b}HVAC stands for heating, ventilating and air conditioning.

5.2 Importance factors among categories

The relative importance of the eleven categories are then established as percentages and are shown below:

(1) lateral resistance   23%
(2) ratio of adjacent dimensions   2%
(3) panel squarish   2%
(4) larger panel dimension reasonably short   23%
(5) live load reasonably light   23%
(6) minimizing height   8%
(7) ready ceiling   2%
(8) simplicity of form work   12%

(9) flexibility of partition	1%
(10) easy poke through	2%
(11) vibration of machinery	2%
	100%

6. DEVELOPMENT OF RULES

The following rules are developed according to the criteria included in the categories secified in section 5.1. The importance factor of a choice in a rule is developed by multiplying the corresponding importance factor of the choice within a category listed in section 5.1 and the relative importance factor of that category specified in section 5.2.

RULE NUMBER: 1

IF:

    Lateral load carried by floor is zero

THEN:

        slab on beams — probability=23/100
  and one-way joist — probability=23/100
  and flat plate   — probability=23/100
  and flat slab   — probability=23/100
  and waffle slab   — probability=23/100

RULE NUMBER: 2

IF:

    Lateral load carried by floor is wind & US seismic zone 0

THEN:

        slab on beams — probability=21/100
  and one-way joist — probability=23/100
  and flat plate   — probability= 9/100
  and flat slab   — probability=12/100
  and waffle slab   — probability=17/100

RULE NUMBER: 3

IF:

    Lateral load carried by floor is wind & US seismic zone 1

THEN:

        slab on beams — probability=21/100
  and one-way joist — probability=23/100
  and flat plate   — probability= 9/100
  and flat slab   — probability=10/100
  and waffle slab   — probability=17/100

RULE NUMBER: 4

IF:
    Lateral load carried by floor is wind & US seismic zone 2

THEN:
        slab on beams — probability=21/100
   and one-way joist — probability=23/100
   and flat plate   — probability= 8/100
   and flat slab   — probability= 8/100
   and waffle slab   — probability=14/100

RULE NUMBER: 5

IF:
    Lateral load carried by floor is wind & US seismic zone 3

THEN:
        slab on beams — probability=21/100
   and one-way joist — probability=23/100
   and flat plate   — probability= 0/100
   and flat slab   — probability= 0/100
   and waffle slab   — probability= 0/100

RULE NUMBER: 6

IF:
    Lateral load carried by floor is wind & US seismic zone 4

THEN:
        slab on beams — probability=21/100
   and one-way joist — probability=23/100
   and flat plate   — probability= 0/100
   and flat slab   — probability= 0/100
   and waffle slab   — probability= 0/100

RULE NUMBER: 7

IF:
    approximate ratio of adajacent spans in same direction is 1.00

THEN:
        slab on beams — probability=2/100
   and one-way joist — probability=2/100
   and flat plate   — probability=3/100
   and flat slab   — probability=3/100
   and waffle slab   — probability=3/100

RULE NUMBER: 8

IF:

    approximate ratio of adajacent spans in same direction is 1.33

THEN:

        slab on beams — probability=2/100
  and one-way joist — probability=2/100
  and flat plate   — probability=2/100
  and flat slab   — probability=2/100
  and waffle slab   — probability=2/100

RULE NUMBER: 9

IF:

    approximate ratio of adajacent spans in same direction is 2.00

THEN:

        slab on beams — probability=2/100
  and one-way joist — probability=2/100
  and flat plate   — probability=1/100
  and flat slab   — probability=1/100
  and waffle slab   — probability=1/100

RULE NUMBER: 10

IF:

    long:short ratio of panels is 1.0

THEN:

        slab on beams — probability=2/100
  and one-way joist — probability=1/100
  and flat plate   — probability=3/100
  and flat slab   — probability=3/100
  and waffle slab   — probability=3/100

RULE NUMBER: 11

IF:

    long:short ratio of panels is 1.1

THEN:

        slab on beams — probability=3/100
  and one-way joist — probability=1/100
  and flat plate   — probability=2/100
  and flat slab   — probability=2/100
  and waffle slab   — probability=2/100

RULE NUMBER: 12

IF:
    long:short ratio of panels is 1.2

THEN:
        slab on beams — probability=3/100
   and one-way joist — probability=1/100
   and flat plate   — probability=2/100
   and flat slab   — probability=2/100
   and waffle slab   — probability=2/100

RULE NUMBER: 13

IF:
    long:short ratio of panels is 1.3

THEN:
        slab on beams — probability=3/100
   and one-way joist — probability=2/100
   and flat plate   — probability=1/100
   and flat slab   — probability=1/100
   and waffle slab   — probability=1/100

RULE NUMBER: 14

IF:
    long:short ratio of panels is 1.4

THEN:
        slab on beams — probability=3/100
   and one-way joist — probability=2/100
   and flat plate   — probability=1/100
   and flat slab   — probability=1/100
   and waffle slab   — probability=1/100

RULE NUMBER: 15

IF:
    long:short ratio of panels is 1.5

THEN:
        slab on beams — probability=2/100
   and one-way joist — probability=3/100
   and flat plate   — probability=1/100
   and flat slab   — probability=1/100
   and waffle slab   — probability=1/100

RULE NUMBER: 16

IF:
    long:short ratio of panels is 1.6

THEN:
      slab on beams — probability=2/100
 and one-way joist — probability=3/100
 and flat plate — probability=1/100
 and flat slab — probability=1/100
 and waffle slab — probability=1/100

RULE NUMBER: 17

IF:
    long:short ratio of panels is 1.7

THEN:
      slab on beams — probability=2/100
 and one-way joist — probability=3/100
 and flat plate — probability=1/100
 and flat slab — probability=1/100
 and waffle slab — probability=1/100

RULE NUMBER: 18

IF:
    long:short ratio of panels is 1.8

THEN:
      slab on beams — probability=2/100
 and one-way joist — probability=3/100
 and flat plate — probability=0/100
 and flat slab — probability=0/100
 and waffle slab — probability=0/100

RULE NUMBER: 19

IF:
    long:short ratio of panels is 1.9

THEN:
      slab on beams — probability=2/100
 and one-way joist — probability=3/100
 and flat plate — probability=0/100
 and flat slab — probability=0/100
 and waffle slab — probability=0/100

RULE NUMBER: 20

IF:
    long:short ratio of panels is 2.0

THEN:
  slab on beams — probability=2/100
 and one-way joist — probability=3/100
 and flat plate   — probability=0/100
 and flat slab   — probability=0/100
 and waffle slab — probability=0/100

RULE NUMBER: 21

IF:
  longer panel dimension is less than 13.5 ft (4.1 m)

THEN:
  slab on beams — probability=14/100
 and one-way joist — probability=12/100
 and flat plate   — probability=23/100
 and flat slab   — probability= 7/100
 and waffle slab — probability= 5/100

RULE NUMBER: 22

IF:
  longer panel dimension is less than 13.5–16.5 ft (4.1–5.0 m)

THEN:
  slab on beams — probability=17/100
 and one-way joist — probability=16/100
 and flat plate   — probability=23/100
 and flat slab   — probability= 8/100
 and waffle slab — probability= 6/100

RULE NUMBER: 23

IF:
  longer panel dimension is 16.5–19.0 ft (5.0–5.8 m)

THEN:
  slab on beams — probability=20/100
 and one-way joist — probability=21/100
 and flat plate   — probability=23/100
 and flat slab   — probability=12/100
 and waffle slab — probability= 8/100

RULE NUMBER: 24

IF:
  longer panel dimension is 19.0–21.8 ft (5.8–6.6 m)

THEN:
  slab on beams — probability=21/100
 and one-way joist — probability=23/100
 and flat plate   — probability=23/100
 and flat slab   — probability=21/100
 and waffle slab — probability=17/100

SELECTION OF SLAB TYPE FOR MULTISTORY BUILDINGS

RULE NUMBER: 25

IF:
    longer panel dimension is 21.8–24.5 ft (6.6–7.5 m)

THEN:
    slab on beams — probability=21/100
  and one-way joist — probability=23/100
  and flat plate — probability=18/100
  and flat slab — probability=23/100
  and waffle slab — probability=22/100

RULE NUMBER: 26

IF:
    longer panel dimension is 24.5–27.5 ft (7.5–8.4 m)

THEN:
    slab on beams — probability=21/100
  and one-way joist — probability=23/100
  and flat plate — probability=15/100
  and flat slab — probability=21/100
  and waffle slab — probability=23/100

RULE NUMBER: 27

IF:
    longer panel dimension is greater than 27.3 ft (8.4 m)

THEN:
    slab on beams — probability=21/100
  and one-way joist — probability=23/100
  and flat plate — probability= 0/100
  and flat slab — probability= 6/100
  and waffle slab — probability=14/100

RULE NUMBER: 28

IF:
    live load on the floor is less than 50 lbf ft^{-2} (2.39) kPa)

THEN:
    slab on beams — probability=14/100
  and one-way joist — probability=14/100
  and flat plate — probability=23/100
  and flat slab — probability= 5/100
  and waffle slab — probability= 2/100

RULE NUMBER: 29

IF:
    live load on the floor is 50.0–60.0 lbf ft^{-2} (2.39–2.87 kPa)

THEN:
        slab on beams — probability=16/100
  and one-way joist — probability=16/100
  and flat plate   — probability=23/100
  and flat slab   — probability= 6/100
  and waffle slab — probability= 3/100

RULE NUMBER: 30

IF:
    live load on the floor is 60.0–70.0 lbf ft^{-2} (2.87–3.35 kPa)

THEN:
        slab on beams — probability=18/100
  and one-way joist — probability=17/100
  and flat plate   — probability=23/100
  and flat slab   — probability=12/100
  and waffle slab — probability= 9/100

RULE NUMBER: 31

IF:
    live load on the floor is 70.0–80.0 lbf ft^{-2} (3.35–3.83 kPa)

THEN:
        slab on beams — probability=21/100
  and one-way joist — probability=18/100
  and flat plate   — probability=21/100
  and flat slab   — probability=23/100
  and waffle slab — probability=18/100

RULE NUMBER: 32

IF:
    live load on the floor is 80.0–90.0 lbf ft^{-2} (3.83–4.31 kPa)

THEN:
        slab on beams — probability=23/100
  and one-way joist — probability=23/100
  and flat plate   — probability=16/100
  and flat slab   — probability=22/100
  and waffle slab — probability=23/100

RULE NUMBER: 33

IF:
    live load on the floor is 90.0–100.0 lbf ft^{-2} (4.31–4.79 kPa)

THEN:
>   slab on beams   — probability=21/100
> and one-way joist   — probability=23/100
> and flat plate   — probability=14/100
> and flat slab   — probability=20/100
> and waffle slab   — probability=23/100

RULE NUMBER: 34

IF:
> live load on the floor is 100.0–110.0 lbf ft^{-2} (4.79–5.27 kPa)

THEN:
>   slab on beams   — probability=18/100
> and one-way joist   — probability=23/100
> and flat plate   — probability= 9/100
> and flat slab   — probability=14/100
> and waffle slab   — probability=23/100

RULE NUMBER: 35

IF:
> live load on the floor is 110.0–120.0 lbf ft^{-2} (5.27–5.75 kPa)

THEN:
>   slab on beams   — probability=16/100
> and one-way joist   — probability=23/100
> and flat plate   — probability= 7/100
> and flat slab   — probability= 9/100
> and waffle slab   — probability=23/100

RULE NUMBER: 36

IF:
> live load on the floor is 120.0–130.0 lbf ft^{-2} (5.75–6.22 kPa)

THEN:
>   slab on beams   — probability=14/100
> and one-way joist   — probability=23/100
> and flat plate   — probability= 0/100
> and flat slab   — probability= 2/100
> and waffle slab   — probability=23/100

RULE NUMBER: 37

IF:
> live load on the floor is greater than 130.0 lbf ft^{-2} (6.22 kPa)

THEN:
>   slab on beams   — probability=12/100
> and one-way joist   — probability=23/100
> and flat plate   — probability= 0/100
> and flat slab   — probability= 0/100
> and waffle slab   — probability=23/100

RULE NUMBER: 38

IF:

    Minimizing building height is important

THEN:

        slab on beams — probability=3/100
   and one-way joist — probability=4/100
   and flat plate   — probability=8/100
   and flat slab   — probability=6/100
   and waffle slab   — probability=4/100

NOTE

Minimizing building height is important if:
(1) local code has height limitation,
(2) reduction of building cladding cost is important,
(3) reduction of power cable cost is important,
(4) reduction of telephone cable cost is important,
(5) reduction of HVAC installation cost is important,
(6) reduction of heating and cooling cost is important,
(7) reduction of window cleaning cost is important.

RULE NUMBER: 39

IF:

    minimizing building height is not important

THEN:

        slab on beams — probability=8/100
   and one-way joist — probability=8/100
   and flat plate   — probability=8/100
   and flat slab   — probability=8/100
   and waffle slab   — probability=8/100

NOTE

Minimizing building height is important if:
(1) local code has height limitation,
(2) reduction of building cladding cost is important,
(3) reduction of power cable cost is important,
(4) reduction of telephone cable cost is important,
(5) reduction of HVAC installation cost is important,
(6) reduction of heating and cooling cost is important,
(7) reduction of window cleaning cost is important.

RULE NUMBER: 40

IF:

    a ready ceiling is important

THEN:

        slab on beams — probability=1/100
   and one-way joist — probability=1/100
   and flat plate   — probability=2/100
   and flat slab   — probability=1/100
   and waffle slab   — probability=1/100

NOTE

Hung ceiling is not required if the floor system provides a smooth surface. Hung ceiling increases the cost of the construction and reduces the available headroom.

RULE NUMBER: 41

IF:

    a ready ceiling is not important

THEN:

    slab on beams — probability=2/100
and one-way joist — probability=2/100
and flat plate — probability=2/100
and flat slab — probability=2/100
and waffle slab — probability=2/100

NOTE
Hung ceiling is not required if the floor system provides a smooth surface. Hung ceiling increases the cost of the construction and reduces the available headroom.

RULE NUMBER: 42

IF:

    simplicity of form work is important

THEN:

    slab on beams — probability= 6/100
and one-way joist — probability=10/100
and flat plate — probability=12/100
and flat slab — probability= 8/100
and waffle slab — probability= 9/100

NOTE
Construction cost can be reduced substantially if the form work of the floor system is simple, especially in the location where cost of form work to material cost is relatively high.

RULE NUMBER: 43

IF:

    simplicity of form work is not important

THEN:

    slab on beams — probability=12/100
and one-way joist — probability=12/100
and flat plate — probability=12/100
and flat slab — probability=12/100
and waffle slab — probability=12/100

NOTE
Construction cost can be reduced substantially if the form work of the floor system is simple, especially in the location where cost of form work to material cost is relatively high.

RULE NUMBER: 44

IF:

    flexibility of partition is important

THEN:
    slab on beams — probability=0/100
  and one-way joist — probability=0/100
  and flat plate — probability=1/100
  and flat slab — probability=1/100
  and waffle slab — probability=1/100

RULE NUMBER: 45

IF:
    flexibility of partition is not important

THEN:
    slab on beams — probability=1/100
  and one-way joist — probability=1/100
  and flat plate — probability=1/100
  and flat slab — probability=1/100
  and waffle slab — probability=1/100

RULE NUMBER: 46

IF:
    easy poke-through hole for wiring, etc., is important

THEN:
    slab on beams — probability=1/100
  and one-way joist — probability=1/100
  and flat plate — probability=2/100
  and flat slab — probability=1/100
  and waffle slab — probability=1/100

RULE NUMBER: 47

IF:
    flexibility of partition is not important

THEN:
    slab on beams — probability=2/100
  and one-way joist — probability=2/100
  and flat plate — probability=2/100
  and flat slab — probability=2/100
  and waffle slab — probability=2/100

RULE NUMBER: 48

IF:
    installation of vibratory machine is anticipated

THEN:
    slab on beams — probability=3/100
and one-way joist — probability=2/100
and flat plate — probability=1/100
and flat slab — probability=1/100
and waffle slab — probability=2/100

RULE NUMBER: 49

IF:
    installation of vibratory machine is not anticipated

THEN:
    slab on beams — probability=2/100
and one-way joist — probability=2/100
and flat plate — probability=2/100
and flat slab — probability=2/100
and waffle slab — probability=2/100

7. EXECUTION OF THE EXPERT SYSTEM

On the basis of the rules discussed in section 6 and using the information input by the user during execution, the currently developed expert system is capable of evaluating the relative position of each of the five floor systems with their corresponding final scores. These final scores are calculated by adding the importance factors for each choice scored in each of 11 categories and dividing by 11. Therefore, the maximum possible score of a choice can only be 100/11=9. For demonstration purposes, the expert system is executed with the following two scenarios.

Scenario 1

The dialogue between the user and the expert system is shown below.

Questions asked by the expert system		Input by the user
(1) Lateral load carried by floor	is	0
(2) Approximate ratio of adjacent spans in same direction	is	1.0
(3) Long:short ratio of panels	is	1.0
(4) Longer panel dimension	is	Less than 13.5 ft (4.1 m)
(5) Live load on the floor	is	Less than 50 lbf ft^{-2} (2.39 kPa)
(6) Minimizing building height	is	Important
(7) A ready ceiling	is	Important
(8) Simplicity of form work	is	Important
(9) Flexibility of partition	is	Important

178 DESIGN OPTIMIZATION

(10)	Easy poke-through hole for wiring, etc.	is	Important
(11)	Installation of vibratory machine	is	Not anticipated

The relative positions of the five choices with their corresponding final scores are shown below.

Floor type	Relative position	Score out of 9
Flat plate	1	9
Slab on beam	2	6
One-way slab	3	6
Flat slab	4	5
Waffle slab	5	4

Scenario 1

The dialogue between the user and the expert system is shown below.

Questions asked by the expert system		Input by the user
(1) Lateral load carried by floor	is	Wind and US seismic zone 4
(2) Approximate ratio of adjacent spans in same direction	is	1.33
(3) Long:short ratio of panels	is	2.0
(4) Longer panel dimension	is	Greater than 27.3 ft (8.43 m)
(5) Live load on the floor	is	Greater than 130 lbf ft^{-2} (6.22 kPa)
(6) Minimizing building height	is	Not important
(7) A ready ceiling	is	Not important
(8) Simplicity of form work	is	Not important
(9) Flexibility of partition	is	Not important
(10) Easy poke-through hole for wiring, etc.	is	Not important
(11) Installation of vibratory machine	is	Anticipated

The relative positions of the five choices with their corresponding final scores are shown below.

Floor type	Relative position	Score out of 9
One-way slab	1	9
Slab on beam	2	7
Waffle slab	3	6
Flat slab	4	3
Flat plate	5	2

Comparison of the dialogues of these two scenarios reveals the appropriateness of the relative standings of the five floor systems in the scenarios.

8. CONCLUSION

The two scenarios discussed in section 7 clearly demonstrate the effectiveness of the expert system developed in this chapter. Because less expensive microcomputer-based expert system shells are now available, similar expert systems to that presented here can be developed and be used cost effectively.

More research in the area of knowledge-based expert systems is still needed. The availability of basic tools for the development of expert systems should facilitate new research in this field. Experts with long industrial experience do not want to share their hard-earned knowledge with others. Participation of these experts is essential for the development of knowledge-based expert systems which will gain industrial acceptance.

REFERENCES

[1] Das, M. L. and Ghosh, S. K., Knowledge bases for multistory concrete buildings, *Computing in Civil Engineering, Proceedings of the Fourth Conference*, American Society of Civil Engineers, New York, NY, October 1986.

[2] Das, M. L. and Sitaram, V. J., Knowledge base for structural design, *Proceedings of First International Conference on Application of Artificial Intelligence to Engineering Problems*, Southampton University, 1986.

[3] Fenves, S. J. and Mahar, M. L., HI-RISE: an expert system for the primary design of high rise buildings, *Tech. Rep.*, Department of Civil Engineering, Carnegie–Mellon University, Pittsburgh, PA, 1983.

[4] Fintel, M. and Ghosh, S. K., Economics of long-span concrete slab systems for office building — a survey, *Concr. Int.*, **5** February 1983.

[5] Harmon, P. and King, D., *Expert Systems: Artificial Intelligence in Business*, Wiley, New York, 1985.

[6] Sriram, D., Destiny: a knowledge-based approach to integrated structural design, *PhD Thesis*, Civil Engineering Department, Carnegie–Mellon University, Pittsburgh, PA, 1983.

[7] Winston, P. H., *Artificial Intelligence*, Addison-Wesley, Reading, MA, 1977.

10

Consultative expert systems for finite element based analysis and design of structure systems

P. Hajela
University of Florida
J. L. Chen
National Cheng Kung University

1. INTRODUCTION

The finite element method for structural analysis has matured considerably over the last three decades, and is now widely used to obtain solutions to complex engineering problems with arbitrary boundary conditions and applied loads. In recent years, the method has been successfully integrated with non-linear programming strategies in optimization, to deliver a very significant capability for optimum synthesis of structural systems. The level of maturity in this field is clearly indicated by the recent inclusion of optimization software in commercially available finite element programs [1,2]. Use of such techniques has been embraced by both the automotive and the aerospace industries [3,4], albeit with some reservations.

These reservations are not totally without basis. Most realistic structural design problems represented by discrete finite element models have a very large number of degrees of freedom. The computational cost associated with the solution of such mathematical models is correspondingly large, and becomes quite prohibitive if the solution process is performed repetitively, as is typically required by non-linear programming algorithms employed for optimum design. Issues pertaining to an efficient implementation of optimization methods in structural design have been studied extensively in several research publications [5–7]. Several useful techniques have emerged which significantly enhance the efficiency of the process. However, these methods can only be adopted in broad-based industry design practice if presented in a form that is comprehensible to a user who is not an optimization expert.

Despite significant advances, current computer-aided design (CAD) capabilities for structural systems are largely restricted to geometric modelling and basic algorithmic analysis of proposed models. Problems encountered in large-scale optimum synthesis have a broader scope, and require a level of judgement of the designer; such judgement is generally derived from extensive experience with similar past problems. The present chapter describes an approach in which the meritorious graphical display and database management features of a CAD system are efficiently incorporated into the knowledge acquisition and representation features of an expert system.

Several publications have explored the idea of applying knowledge-based expert systems to assist in finite element modelling [8,9]. The generation of a finite element model includes selection of element types, node selection, mesh generation, load idealizations, and specification of boundary conditions. Without significant prior experience in finite element modelling and in in-depth understanding of the capability of each element, mismodelling can occur frequently [10]. Furthermore, if the finite element model is to be used in optimum synthesis, it is necessary to emphasize the link between modelling for analysis and for efficient optimum design. The quality of the design model and the choice of a suitable optimization algorithm with its proper parameters has a sizeable influence on the results of the structural optimization. The formulation of the optimum design model generally entails the selection of design variables, formulation of constraints and objective function, and a choice of the optimization strategy.

Prior attempts in the application of artificial intelligence (AI) techniques in the field of optimum design are documented in references [11–14]. The central theme of all such efforts has resided in the use of engineering knowledge and AI methods to improve the efficiency of the optimization process. The EXADS system [15] was developed as a consultant to assist in the selection of an optimization algorithm from the ADS general-purpose optimization program [16]. Another expert system IDEAS [17] is under development to assist in the overall non-linear optimization problem.

The present chapter provides an overview of an expert system OPSYN (optimum synthesis) for optimum structural synthesis, with particular emphasis on the significance of the development environment for such systems. The framework for this development is an inference engine with both forward and backward reasoning capabilities and a detailed explanation facility. The knowledge base consists of rules for finite element modelling, optimum design modelling, and selection of optimization strategies and parameters. The finite element program EAL [18] and the ADS general-purpose optimization program are selected to perform finite element analysis and optimization respectively. This chapter also focuses on underscoring the usefulness of integrating expert systems into the CAD environment, and uses the significant data structure and graphics capabilities of the latter in the knowledge acquisition and representation process. Subsequent sections of this chapter describe the extent of the knowledge base in greater detail. Further, the organization of the knowledge base is discussed in the context of the present inference facility.

2. FINITE ELEMENT MODELLING

The success of a finite element based analysis is generally dependent on the quality of the discrete model. Finite element modelling is recognized as a skill that requires a degree of engineering judgement. It is acquired through exposure to a range of problems and through the experience of other users. Generating the finite element model usually involves a process of idealization and discretization. A proper idealization of the structural problem is a significant step in finite element modelling, and pertains to selecting the types of elements. An element is typically characterized by its behavioural properties (i.e. truss, beam, membrane, plate, etc.), order (linear, quadratic), or shape (linear, triangular, quadrilateral).

Selection of the best element for the problem at hand requires a considerable knowledge of the characteristics of each element. It also requires an insight into the physical problem, including recognition of failure modes that are most likely for a given problem and affected by the shape of the problem domain, the prescribed set of loads, and the boundary conditions. Unfortunately, few guidelines for choosing optimum elements can be established, as the type of element that yields good accuracy with low computational requirements is highly problem dependent. In addition, there is a growing concern about the reliability of commercially available finite element codes. The recent organization of an agency in the UK to propose test problems for finite element code standardization [19] is a testimony to this concern. Results of such studies have been recently published [20], and it is expected that these and similar results will provide more general guidelines to assist in finite element modelling.

Discretization of the structure system involves decisions on the number, shape, size, and distribution of finite elements in the problem domain. Proper specification of node locations and elements shapes is essential in order that the discretized domain and the actual domain are as close as possible. Due consideration must also be given to an accurate representation of concentrated loads, distributed loads, and proper handling of geometric and material discontinuities, including cutouts and re-entrant corners. The domain must further be discretized into sufficiently small elements to resolve any spatial variations in the solution. In order to do this effectively, an *a priori* understanding of the solution and associated computational requirements is necessary. Typical rules of thumb dicatate that due consideration be given to element aspect ratio, element distortion, any symmetry or lack thereof of loading and geometry in the problem domain, and the node numbering sequence. The latter is particularly significant from a computational efficiency standpoint. There has been an increased focus on automated mesh generation and mesh refinement techniques to improve the mesh layout. However, these techniques are not always practically feasible, and one must use prior experience in this problem. It is clear from the foregoing discussion that a considerable amount of experience and engineering judgement is needed for establishing an optimum finite element model. The development of a knowledge base to provide online assistance in this task is seen as a step to bringing together the accumulated knowledge of experienced users in finite element modelling.

3. MODELLING FOR DESIGN AND OPTIMIZATION

The use of numerical mathematical programming techniques in conjunction with a finite element based analysis provides a viable approach for the optimum sizing of complex structural configurations. The mathematical formulation of a structural optimization problem can be written as follows:

minimize: $W(d)$
subject to: $g_j(d)<0$
$\qquad\qquad\qquad j=1,2,\ldots,m$
$\qquad\quad h_k(d)=0 \quad k=1,2,\ldots,p$

where W is the objective function and is typically the structural weight, d is a vector of design variables, and q and h are inequality and equality constraints respectively on the response. Realistic structures may have hundreds of design variables and thousands of implicit non-linear constraints. Most efficient non-linear programming algorithms are gradient based, and require sensitivity of the objective and constraint functions with respect to the design variables. If applied directly, this approach can become prohibitively expensive to implement in a realistic problem. To circumvent the problem and to improve the efficiency of the process, several approximation techniques have been devised. These include reduction in the number of design variables by design variable linking, reduction of constraints by using efficient constraint representations, limiting the number of finite element analyses through approximate analysis methods, and by decomposing the problem into a sequence of smaller optimization problems.

At the very outset, the designer must consider all available means to establish an efficient design model. The possibility of generating a simplified analysis model for use in a repetitive optimization cycle, dimensionality reduction by invoking approximation concepts, and use of problem decomposition approaches must all be considered. This step is followed by the actual formulation of the optimum design model which includes the specification of the kind, number, and distribution of design variables, identifying the load conditions and constraints to be considered during the optimization, and selection of a suitable objective during the optimization, and selection of a suitable objective function for the problem. The proper choice of design variables is very critical as the objective function and constraints may vary linearly or near-linearly with a particular choice of design variables, and may exhibit significant non-linear behaviour with respect to another choice. A linear or near-linear behaviour is desirable as it eases the computational cost of obtaining gradients. In some cases it is actually better to not use member-sizing variables as the design variables for optimization, but rather a combination of function or such variables for which the behavioural response is linear. Such variables are termed intermediate variables and are problem dependent.

After the design model is formulated, the designer also needs to determine a suitable optimization strategy and an approach to computing behavioural sensitivity so as to enhance the efficiency of the overall process. The available optimization strategies include fully stressed design, optimality criterion methods, mathematical programming, and a dual design space approach. Each of these methods has its advantages [21], and its application is influenced by the characteristics of the problem. In a similar manner, the choice of a methodology to determine the behavioural sensitivity from among the finite difference method, design space approach, behaviour space approach, and virtual load technique is also problem dependent.

Most of the currently available optimization alghorithms used in the solution of optimal design problems in engineering require a certain level of expertise of the user. Some such expertise can be built into the knowledge base of a general expert system for optimum structural synthesis. In particular, a user is confronted with monitoring the execution for numerical error diagnostics, adaptively selecting and switching the optimization algorithm depending on the local characteristics of the design problem, selection of pertinent parameters for the algorithm of choice, identification of important design variables and critical constraints, and recognition of an ill-defined problem space. Several of these tasks require significant quantitative information about the problem space and this necessitates that such an expert system be integrated with conventional algorithmic programs. Further details of such an implementation are available in [22].

4. FRAMEWORK FOR DEVELOPMENT — ARCHITECTURE OF OPSYN

The role of an expert system as an online consultant for the task shown schematically in Fig. 10.1. Each of the tasks of finite element modelling, optimum design modelling and use of an optimization algorithm in conjunction with analysis is regulated to a modular knowledge base, and embedded in the inference environment. The knowledge base consists of domain facts and the heuristic knowledge associated with the problem. An IF(condition)–THEN(action) rule-based knowledge representation scheme is used in the present system. This is a popular knowledge representation scheme wherein the action part is executed if the condition part is matched with available facts.

The structure of the OPSYN system is illustrated in Fig. 10.2. It contains a knowledge base, a backward and forward reasoning inference engine with an explanation facility, a knowledge acquisition facility, and an input–output facility that includes a knowledge base editor and a graphical display capability. An overview of each component of this system is provided in the following sections.

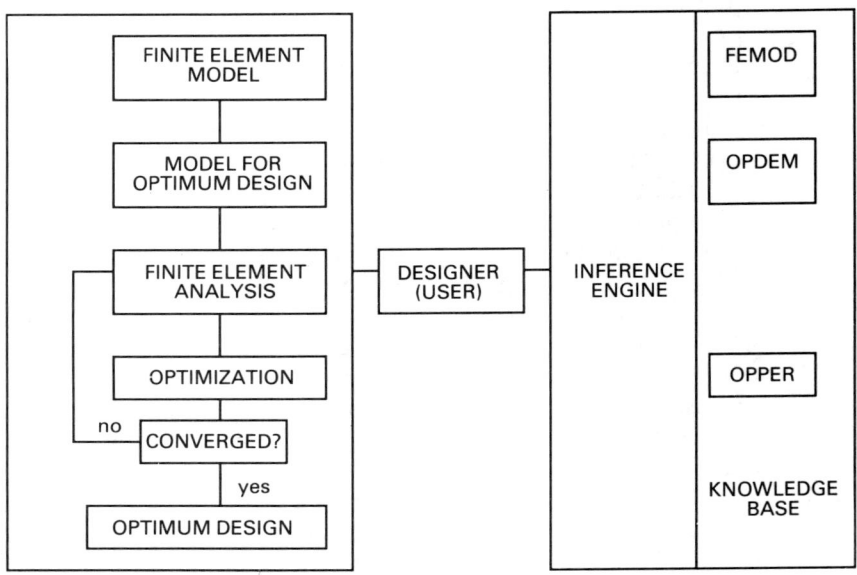

Fig. 10.1 — The role of an expert system in optimum modelling for structural analysis.

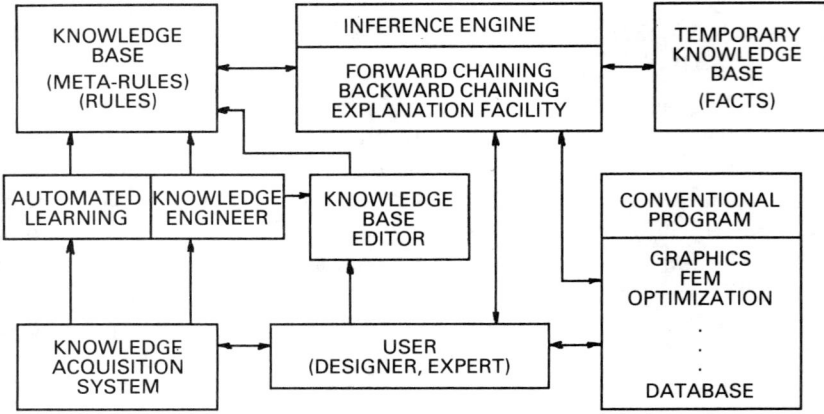

Fig. 10.2 — The structure of the OPSYN system.

4.1 Knowledge base organization

The organization of the knowledge base for optimum structural synthesis is shown in Fig. 10.3. As described earlier, the knowledge base is separated

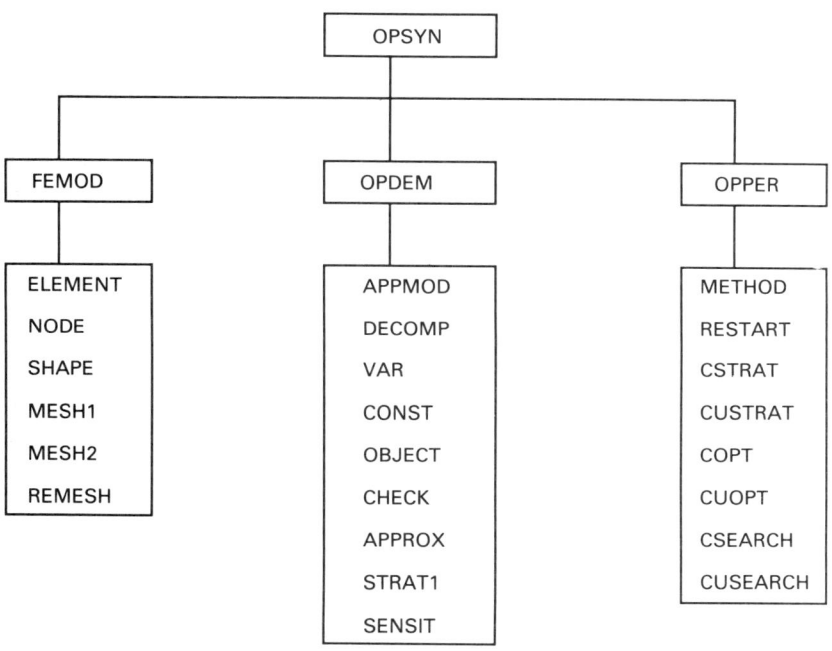

Fig. 10.3 — Organization of the knowledge base for optimum structural synthesis.

into these modules according to the task in different domains. Such a division promotes the modularity of the system and assists in speeding up the reasoning process. The module FEMOD contains rules for finite element modelling. Rules for optimum design modelling and selection of optimization strategies and parameters are grouped into the OPDEM module and the OPPER module respectively. In each of these modules, the knowledge base is further separated into different files. Such an arrangement was necessary for applying different inference reasoning techniques to various aspects of the problem and to make the inference process more efficient.

The rules for finite element modelling include information pertaining to location of nodes, node numbering, mesh generation, mesh refinement, element selection, and guidelines to eliminate element distortion. The rules for optimum design modelling are primarily intended to create a numerically efficient analysis model to be used repetitively in the optimization process. This includes rules for selection of design variables, constraints, approximation techniques, and strategies for sensitivity analysis. A third set of rules are

188 DESIGN OPTIMIZATION

proposed to enhance optimization performance, and these primarily assist the user in selecting optimization strategies for the design problem. This includes rules for both unconstrained and constrained optimization problems, and rules for algorithm switching in the event that the initial selection is unsatisfactory. Additional details about these rule bases and their applications are available in references [23–26].

4.2 INFER inference facility

A domain-independent, rule-based inference engine INFER was developed especially for applications in engineering analysis and design. It includes forward, backward, and exhaustive backward reasoning, and an extensive explanation facility. Incomplete knowledge can be handled by associating confidence levels with the facts. In addition, there is a special feature to interface with conventional algorithmic programs. The inference environment also has a knowledge acquisition system and a knowledge base editor to assist in the knowledge elicitation process.

INFER was developed to be operational in a CAD environment. It is written in FORTRAN-77, a feature that allows easy access to existing graphics subroutines. Currently, INFER is operational on a VAX 11-750 system. A Tektronix 4107 graphics terminal serves as the user interface.

The flowchart for the INFER inference engine is shown in Fig. 10.4. At

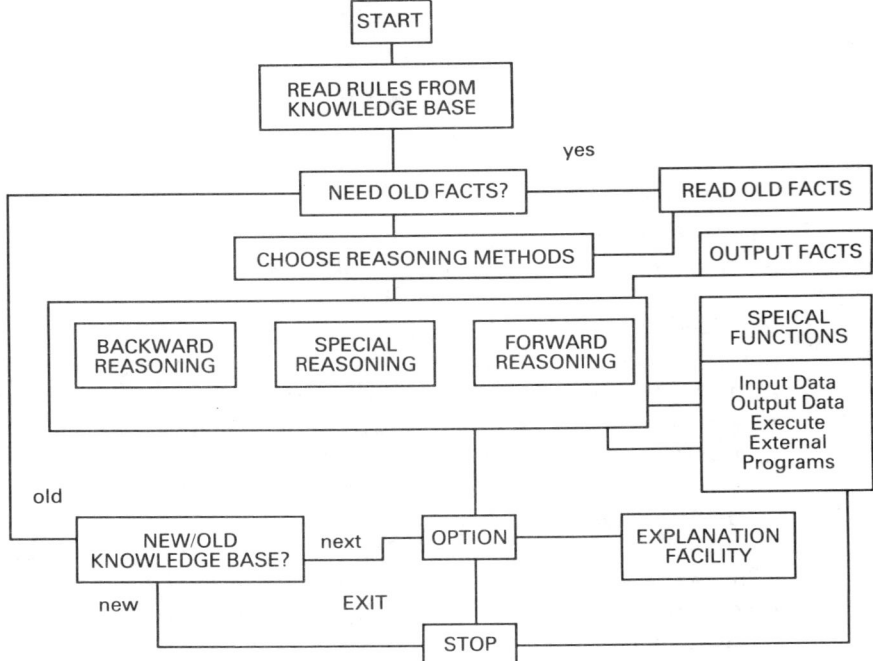

Fig. 10.4 — Flowchart for the INFER inference engine.

the onset of the reasoning process, the rules in the knowledge base are first read. As explained in a later section, these rules are both in a text and a graphical format. The next step, if necessary, is to read the existing facts from the temporary facts database. These may be additional facts available at the outset of the inference process or facts that may have been established in a prior execution. Choosing a suitable inference reasoning technique is the third step, and INFER allows the user to make an appropriate selection. The available techniques are forward, backward, or special (exhaustive backward) reasoning. The fourth step shown in the flowchart is the inference reasoning process itself. If a conclusion is confirmed and it contains a function symbol in the statement, then the program executes this function. After the reasoning process, the user has the option of seeking an explanation, performing another task, or terminating execution of the program.

The INFER environment has a knowledge base editor facility to help the knowledge engineer in editing and modifying the rules in the knowledge base. This facility both reduces the expert system development time and shortens the learning time for a new user. The knowledge base editor program KBSED can perform a variety of tasks, such as creating new rules, modifying old rules, deleting old rules, changing rule order, displaying rules, changing knowledge base type, and updating the knowledge base. The task of modifying old rules includes changing confidence levels, adding conditions or actions, deleting condition of actions, and modifying conditions or actions. KBSED also has the capability to create and modify graphical data.

4.3 Representation of knowledge

In OPSYN, knowledge is represented in the IF-THEN rule-based scheme. A confidence level value is associated with each rule. Further, rules are divided into two categories, meta-rules and rules. An example of a meta-rule in the module dealing with optimum design modelling, OPDEM, is as follows:

THIS IS RULE 1

CONFIDENCE LEVEL = 1.0

IF

    This problem is at the first step of optimum design modelling.

THEN

    You need to consider whether the computational time and/or accuracy are important for this problem. You should run the APPMOD knowledge base by using backward chaining reasoning.

AND

    ∗∗∗COPY APPMOD.KBS;1 INPUT.DAT;3

The '∗∗∗' symbol is a special function symbol in the INFER inference engine. When encountered during a program execution, it is interpreted as a flag to create a system executable command file which essentially contains the statement following the symbol in the rule. This command file is also executed. Another example of a rule in the RESTART knowledge base is as follows:

THIS IS RULE 2

CONFIDENCE LEVEL = 1.0

IF

Treatment of the problem as a constrained problem was advised.

AND

You have already modified some parameters (i.e., push-off factor, constraint tolerance, etc.).

THEN

You should switch to methods that treat the constrained problem as an unconstrained problem. The extended interior penalty function strategy (ISTRAT-2,3,4) is advised.

A principal drawback of current knowledge representation schemes in most expert systems is the frequency with which misrepresentation of knowledge can occur, particularly of text alone is used as mode of communication. In order to overcome this limitation, a graphical display function was developed within the INFER environment. Function symbols (in this case '&&&') were appended to rule and actions to provide a graphical explanation of the text on the display terminal. An example of rule with an additional graphical display feature is as follows:

IF

&&& This problem has distributed loads.

THEN

&&& Choose nodes at all points where distributed loads change, and subdivision lines through these points are normal to the boundary.

The actual graphical display of this rule and action appears in Fig. 10.5.

Fig. 10.5 — Graphical display of a rule and action.

Another example of a condition in the current MESH2 knowledge base is as follows:

&&& This problem has a quadrilateral domain or can be considered a combination of several quadrilateral domains:
Note that curved boundaries can be represented by a combination of linear segments, and the analysis domain may indeed be quadrilateral.

The actual graphical display of this condition on the display terminal is shown in Fig. 10.6. The addition of this feature is recognized as playing a

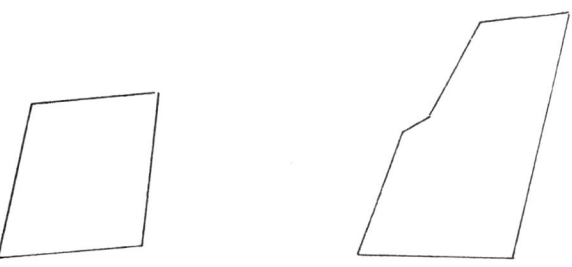

Fig. 10.6 — Graphical display of a rule and action.

significant role in eliminating misrepresentation of rules and conditions–actions.

4.4 Knowledge acquisition in a CAD environment

As in previous expert system development, the process of knowledge acquisition was recognized as the most difficult and challenging task of the proposed effort. A conventional approach was adopted, wherein a domain expert was tapped as the principal source of knowledge for the problem. Prior experiences in solving problems in optimum structural synthesis were presented and discussed with the expert to validate their inclusion. A secondary source of knowledge was the documentation available in the literature. These two modes of knowledge acquisition were used extensively in the initial phase of expert system development.

The CAD environment in which the expert system was developed makes available a significant capability for knowledge elicitation from the domain expert. Such a system provides an online capability for dynamically updating the knowledge base over a period of time. The use of a CAD environment as a tool to elicit domain knowledge falls in the broad category of methods of protocol analysis [27]. In the present implementation, several experts are required to perform modelling for analysis and design of given problems. Their decisions at various stages of the modelling process, and the confidence levels that they assign each response, are stored in a file for postprocessing. The knowledge acquisition system includes two major components. These are the ACQU2 program to elicit knowledge from the domain experts, and a second program RULE to evaluate the knowledge and to formulate rules. The relationship between these programs is illustrated schematically in Fig. 10.7.

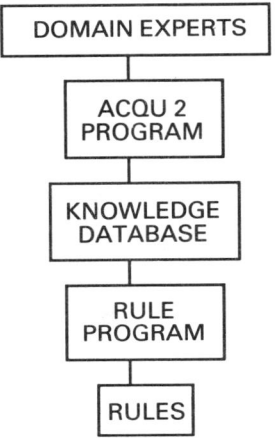

Fig. 10.7 — Relationship between the ACQU2 and RULE programs.

When using the CAD-based ACQU2 program in a knowledge acquisition mode, the expert makes a task-related decision and must also provide a confidence level for the decision. The expert can either select from a set of predefined responses or provide an alternative response. This information is recorded in an IF–THEN rule format in a temporary storage file. The knowledge engineer then executes the program RULE to analyse this information. The items included in such a post-acquisition analysis include the number of domain experts that supported a rule, the average value of confidence level that the experts assigned a rule, and the distribution of confidence levels associated with this rule. On the basis of this information, the knowledge engineer can select the best rule for a hypothesis. The RULE program also provides information on the difference of each expert's response from the average of those supplied by other experts. This allows the knowledge engineer to scan each domain expert's response to evaluate whether a possible skew exists in that expert's assignment of confidence levels. A suitable weighting factor may then be added to that expert's response. This option prevents improper responses from biasing or distorting the knowledge base. An additional feature of the RULE program is to

Ch. 10] ANALYSIS AND DESIGN OF STRUCTURE SYSTEMS   193

allow the domain expert to modify old records as more information is available, and to update the existing knowledge base.

An illustration of the use of such a knowledge acquisition system is available in the finite element modelling of an edge-loaded square flat plate with a hole in the structural domain. The hole can be a circle, an ellipse, or a rectangle (Fig. 10.8). The position and dimension of this hole can be

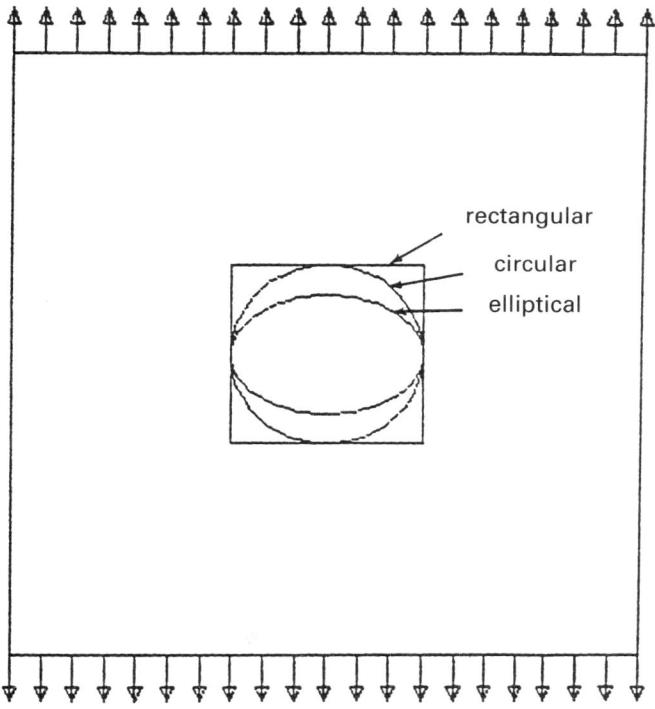

Fig. 10.8 — Edge-loaded square with a hole.

changed and defined by the designer in order to reflect every situation during the finite element modelling process. During the process of using the CAD-based ACQU2 program to elicit finite element modelling knowledge, the nodes in the model can be generated by a direct entry of the nodal coordinates. A light-pen or cursor may also be used for this task. After a few key nodes are entered, the others may be generated by specifying interpolatory functions between these key nodes and the desired nodes. The mesh can be generated by identifying individual element node numbers or by specifying the node numbers on the desired subdivision lines. The following are examples of the decisions made by the domain expert and recorded in the temporary storage file.

CONFIDENCE LEVEL = 1.0

IF

  PLANE STRESS PROBLEM

THEN

  CHOOSE E41 QUADRILATERAL MEMBRANE ELEMENTS

CONFIDENCE LEVEL = 0.9

IF

  BOUNDARY POSITION

AND

  THE BOUNDARIES CHANGE SHARPLY

THEN

  CHOOSE NODE AT 1 (250.0, 250.0)
  CHOOSE NODE AT 2 (1000.0. 1000.0)

The records stored in the database can be viewed as a three-dimensional decision table, as shown in Fig. 10.9. The execution of the RULE program

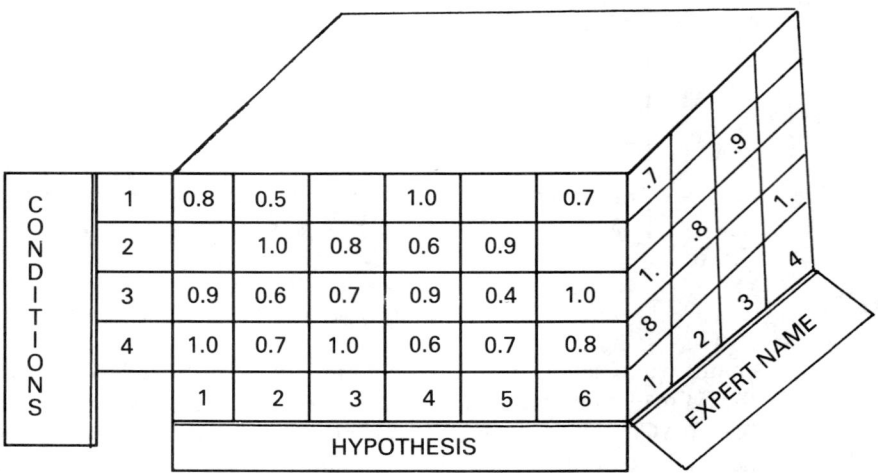

Fig. 10.9 — Records stored in a database derived from the decisions of a domain expert viewed as a three-dimensional decision table.

to analyse the knowledge in the database file results in a printout of lists of pertinent information for each rule. For example,

TOTAL VALUES OF CONFIDENCE LEVEL = 8.5

NUMBER OF EXPERT PERSONS = 10
AVERAGE CONFIDENCE LEVEL = 0.850

The distribution of confidence levels for a rule is shown in Table 10.1. On

Table 10.1 — Typical distribution of confidence levels for a rule

Range of confidence levels for a given rule	Number of experts supporting the this level of confidence
1.0–0.9	6
0.9–0.8	2
0.8–0.7	1
0.7–0.6	0
0.6–0.5	1
0.5–0.4	0
0.4–0.3	0
0.3–0.2	0
0.2–0.0	0

the basis of this information for each rule, the knowledge engineer can select the best rule for a specified hypothesis. The results obtained in the initial implementation and testing of the system vindicate the usefulness of this approach. The vailability of such a knowledge acquisition system can help the knowledge engineer in eliciting knowledge from the domain experts and implementing the knowledge in the knowledge base.

5. ILLUSTRATIVE EXAMPLES

In this section, some single examples are presented to illustrate the use of the OPSYN system. The first example deals with the finite element modelling of an edge-loaded, thin plate with a circular hole, discussed in the previous section. A second example in the domain of finite element modelling is a one-bay, two-storey frame structure with concentrated and distributed loads, as shown in Fig. 10.10. The frame is designed for minimum weight and subject to maximum allowable stresses in the member. The finite element models for these problems are shown in Figs 10.11 and 10.12 respectively. A third example illustrating the task of optimum design modelling of a cantilevered beam with a fixed tip mass is shown in Fig. 10.13. This structure is designed for minimum weight, subject to frequency constraints. Abbreviated session maps of interactive sessions with the expert system are presented in Appendix 1.

6. CONCLUDING REMARKS

An expert system, OPSYN, has been developed to aid a structural designer in the computer-aided optimum design of structural and mechanical

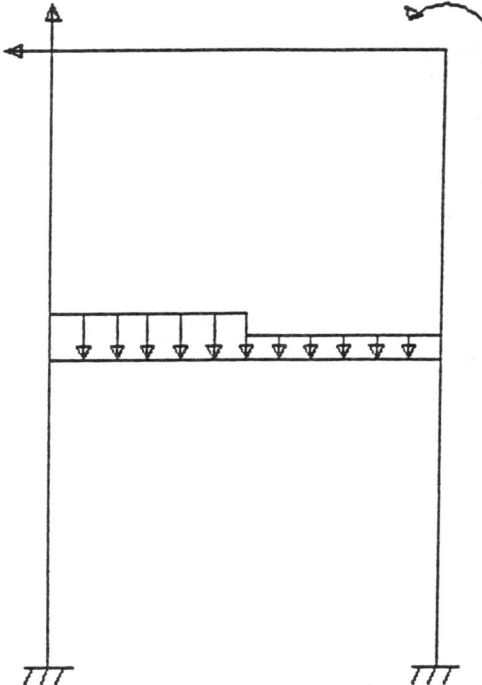

Fig. 10.10 — Example of the use of OPSYN: a one-dimensional, two-story frame structure.

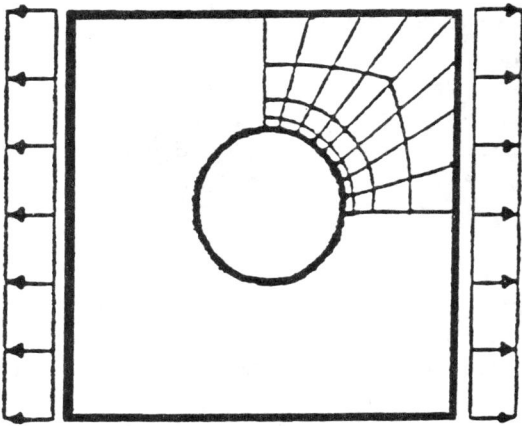

Fig. 10.11 — Example of the use of OPSYN: finite element model for an edge-loaded thin plate (see Fig. 10.8).

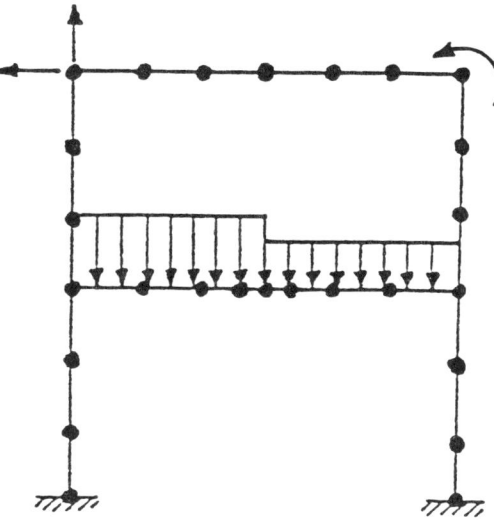

Fig. 10.12 — Example of the use of OPSYN: finite element model for the structure of Fig. 10.10.

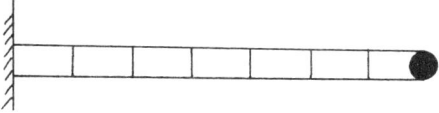

Fig. 10.13 — Example of the use of OPSYN: a cantilevered beam with a fixed tip mass.

systems. The OPSYN expert system can provide interactive assistance in finite element modelling for the finite element program EAL, optimum structural design modelling, and selection of optimization strategies and parameters for the general-purpose optimization programs ADS.

The environment for this development is the INFER inference engine. INFER was especially developed for the building of expert systems for engineering design and analysis and was intended to be operational in a CAD environment. The graphical display capability within the CAD environment was used to represent some engineering knowledge more concisely than the conventional text-based representation. During the OPSYN expert system development, the knowledge-acquisition process was recognized as the most difficult task. An approach using existing CAD capabilities as a tool to elicit knowledge from the domain expert directly was presented. This approach requires several domain experts to perform the tasks on a set of prestructured problems. The responses from domain experts are stored in a file, which can be analyzed to propose rules for the knowledge base.

The preliminary testing and verification of the current approach used in the OPSYM expert system give satisfactory results. It is believed that a system such as OPSYN would enhance the role of structural optimization in the design community. However, a significant limitation of the current OPSYN expert system is that it operates only on the basis of responses provided by the user. Additional development work is necessary on the subject of utilizing the existing capability of communicating with conventional analysis programs in the INFER environment. This can be used to obtain more reliable and quantitative information about the problem domain.

REFERENCES

[1] Wallerstein, D. V., Design enhancement tools in MSC/NASTRAN, *NASA CP 2327*, Part I, April 1984.
[2] Stefenides, E. J., Revised finite element analysis software proffers better design tools, *Des. News*, 100–106, April 7, 1986.
[3] Bennett, J. A. and Botkin, M. E., Automated design for automotive structures, *J. Mech. Des.*, **104**, 799–805, October 1982.
[4] Brama, T., Weight optimization of aircraft structures. In Soares, M. (ed.), *Proceedings of NATO/NASA/NSF/USAF Advance Study Institute on Computer-Aided Optimal Design, Troia, Portugal*, Vol. 3, 1986, pp. 267–277.
[5] Schmit, L. A., and Miura, H., Approximation concepts for efficient structural synthesis, *NASA CR-2552*, 1976.
[6] Vanderplaats, G. N., Miura, H., and Chargin, M., Large scale structural synthesis, *Finite Elem. Anal. Des.*, **1**(3), 117–130, 1985.
[7] Hajela, P., Geometric programming strategies for large-scale structural synthesis, *AIAA J.*, **24**(7), 1986.
[8] Woodward, W. S. and Morris, J. W., Improving productivity in finite element analysis through interactive processing, *Finite Elem. Anal. Des.*, **1**(1), 35–48, 1985.
[9] Fenves, S. J., A framework for a knowledge-based finite element analysis assistant. In Dym, C. L. (ed.), *Applications of Knowledge Based Systems to Engineering Analysis and Design*, ASME, New York, 1985, pp. 1–7.
[10] Smith, G. E., The dangers of CAD, *Mech. Eng.*, 58–64, February 1986.
[11] Arora, J. S. and Baenziger, G., Use of AI in design optimization, *Comp. Methods Appl. Mech.*, **54**(3), 303–323, 1986.
[12] Papalambros, P., Knowledge-based systems in optimal design. In

Soares, M. (ed.), *Proceedings of NATA/NASA/NSF/USAF Advance Study Institute on Computer-Aided Optimum Design, Troia, Portugal*, Vol. 3, 1986, pp. 311–362.

[13] Jha, N. K., Engineering optimization and expert systems, *Proceedings of the 1985 ASME International Computers in Engineering Conference*, Vol. 3, ASME, New York, 1985, pp. 97–101.

[14] Jozwiak, S. F., Application of artificial intelligence notions in structural optimization programs, *Comput. Struct.*, **24**(6), 1009–1013, 1986.

[15] Rogers, J. L., Jr. and Barthelemy, J.-F. M., An expert system for choosing the best combination of options in a general purpose program for automated design synthesis, *Eng. Comput.*, **1**(4), 217–227, 1985.

[16] Vanderplaats, G. N., Sugimoto, H. and Sprague, C. M., ADS-1: a new general-purpose optimization program, *AIAA J.* **32**(10), 1458–1459, October 1984.

[17] Arora, J. S. and Baenziger, G., A non-linear optimization expert system, In Computer Applications in Structural Engineering, Jenkins, D. R. (ed.), ASCE, New York, 1987, pp. 113–125.

[18] Whetstone, D., *EISI–EAL Engineering Analysis Language Reference Manual — EISI–EAL System Level 2091*, Engineering Information Systems, San Jose, CA, July 1983.

[19] Mair, W. M., The objectives of the National Agency for Finite Element Methods and Standards, *Comput. Struct.* **21**(5), 875–879, 1985.

[20] Macneal R. H. and Harder, R. L., A proposed standard set of problems to test finite element accuracy, *Finite Elements Anal. Des.*, **1**(1), 3–20, April 1985.

[21] Kirsch, U., *Optimum Structural Design — Concepts, Methods and Applications*, McGraw-Hill, New York, 1981.

[22] Chen, J. L., and Hajela, P., Integration of algorithmic computations in expert systems, *Proceedings of the 12th IMACS World Congress on Scientific Computation, Paris, July 1988*.

[23] Chen, J. L. and Hajela, P., FEMOD: a consultative expert system for finite element modeling, *Comput. Struct.*, **29**(1), 1988.

[24] Chen, J. L. and Hajela, P., A rule based approach to optimum design modelling. In Jenkins, D. R. (ed.), *Computer Applications in Structural Engineering*, ASCE, New York, 1987, pp. 66–81; accepted for publication in *Comput. Struct.*

[25] Hajela, P. and Chen, J. L., CAD interfaces in the development of expert systems for optimum structural design, *Proceedings of the Second International Conference in Computational Engineering Science, Atlanta, GA, April 1988*, in *Computational Mechanics*, Springer, Berlin, 1988.

[26] Chen, J. L., A knowledge-based expert system consultant for optimum structural synthesis, *PhD. Dissertation*, University of Florida, Gainesville, FL, August 1987.

[27] Hart, A., Knowledge elicitation: issues and methods, *Comput. Aided Des.*, **17**(9), 455–462, November 1985.

APPENDIX 1

Example 1: edge-loaded thin plate with circular hole

$ COPY FEMOD.KBS;1 INPUT.DAT;1
$ RUN INFER

This problem is a plane stress problem.
0.8

PLEASE ANSWER THE FOLLOWING QUESTION WITH A FACTOR RANGING FROM 0.0 (DEFINITE "NO") TO 1.0 (DEFINITE "YES")
??
This problem is such that only inplane loads are applied, and that there are no rotational degrees of freedom permitted at any point of structure modeled by these elements.
0.7

****

THE HYPOTHESIS THAT HAS CERTAINTY 0.900

This problem requires elements E31 or E41. The E31 are triangular membrane elements. The E41 are quadrilatral membrane elements.

⋮

$ RUN INFER

⋮

??
this problem has symmetrical geometry and loading.
1.0

****

THE HYPOTHESIS THAT HAS CERTAINTY 1.000

Symmetry indicates a reduced analysis problem: one or two axes of symmetry would require half or quarter domain. Additional lines of symmetry dictate a corresponding reduction in the analysis domain. Choose subdivision lines along axes of symmetry and nodes on these lines.

⋮

??
The domain has holes or cutouts.
0.9

****

THE HYPOTHESIS THAT HAS CERTAINTY 0.900

You need a refined mesh for the domain in the proximity of the discontinuity. For static analysis problems: choose nodes in the domain around the hole, at least 2 to 5 layers around the hole, the mesh-size-ratio (typically defined as the size of a cell closest to discontinuity to the cell size in the next layer of elements) should be about 0.5. For the narrow domain between the outer and inner boundary, you should also consider a refined mesh.

⋮

$ RUN INFER
   ⋮

??
&&& This problem has a quadrilateral domain or can be considered a combination of several quadrilateral domains: note that curved boundaries can be represented by a combination of linear segments, and the analysis domain may indeed be quadrilateral.
0.9
   ⋮

??
This problem considers only yielding case.
1.0
   ⋮

??
This problem is under only static loading.
1.0

****

THE HYPOTHESIS THAT HAS CERTAINTY 0.900

&&& Wherever possible use quadrilateral elements. Use triangular elements if domain boundary is such that these elements are required or if these elements can be used in conjunction with quadrilateral elements to maintain any inherent symmetry in the problem. Using either quadrilateral or triangular elements, it is important to retain any mathematical symmetry that may exist, that is both loading or support symmetry. Number of elements should be such that there are at least 4 to 8 nodes along the longer side and 3 to 4 nodes along the shorter dimension. For a square domain use at least 4 to 8 nodes along each side.

$ RUN INFER
   ⋮

??
The mesh has been generated for this problem.
1.0

PLEASE ANSWER THE FOLLOWING QUESTION WITH A FACTOR RANGING FROM
0.0 (DEFINITE "NO") TO 1.0 (DEFINITE "YES")
??
You want to check element distortion, aspect ratio, or mode numbering.
0.9

****

THE HYPOTHESIS THAT HAS CERTAINTY 0.900

You should run the SHAPE knowledge base to check distortion, aspect ratio, or node numbering by using special reasoning.
***COPY SHAPE.KBS;1 INPUT.DAT;1
   : :

****

THE HYPOTHESIS THAT HAS CERTAINTY 0.900

For quadrilateral elements: keep interior corner angles near 90 degree and always between 0 and 180 degrees.
For rectangular elements, the sides of the element should be kept parallel to the cooordinate axis in as far as possible.
For triangular elements: the element should be kept as close as possible to an equilateral geometry.
If you think you followed these instructions and still have distortion, then look for elements or nodes that you may have inadvertently omitted.

PLEASE ANSWER THE FOLLOWING QUESTION WITH A FACTOR RANGING FROM
0.0 (DEFINITE "NO") TO 1.0 (DEFINITE "YES")
??
You want to check the aspect ratio for the elements.
1.0

****

THE HYPOTHESIS THAT HAS CERTAINTY 0.900

If displacements are the principal concern, then aspect ratio should be less than 7.0.
If stresses are needed in addition, choose aspect ratio less than 3.0.

PLEASE ANSWER THE FOLLOWING QUESTION WITH A FACTOR RANGING FROM
0.0 (DEFINITE "NO") TO 1.0 (DEFINITE "YES")
??
You wish to look at the rules which would help you arrange node numbers.
0.9

****

THE HYPOTHESIS THAT HAS CERTAINTY 0.900

Smaller bandwidths obtained by numbering the nodes in the direction of the smallest number of subdivision lines.

Example 2: one-bay, two-storey frame with distributed and concentrated loads

$ COPY FEMOD.KBS;1 INPUT.DAT;1
$ RUN INFER

DID YOU ASSIGN THE KNOWLEDGE BASE FILE AS "INPUT" FILES? OR COPY TO "INPUT.DAT" FILE?
1 → YES, 0 → NO
1
    READING RULE

DO YOU WISH TO USE FACTS IN "FACTS.DAT" FILE?
1 → YES, 0 → NO

PLEASE CHOOSE INFERENCE PROCESS?

1 → FORWARD CHAINING (YOU ALREADY KNOW SOME FACTS!)
2 → BACKWARD CHAINING
3 → SPECIAL REASONING (ASK QUESTIONS FOR EVERY RULE!)
2

***BACKWARD CHAINING PROCESS★★★

PLEASE ANSWER THE FOLLOWING QUESTION WITH A FACTOR RANGING FROM
0.0 (DEFINITE "NO") TO 1.0 (DEFINITE "YES")
??
The mesh has not been generated for this problem.
1.0

PLEASE ANSWER THE FOLLOWING QUESTION WITH A FACTOR RANGING FROM
0.0 (DEFINITE "NO") TO 1.0 (DEFINITE "YES")
??
The FEM elements have not been selected for this problem.
1.0

****

THE HYPOTHESIS THAT HAS CERTAINTY 1.000

You should run the ELEMENT knowledge base to choose elements for your problem by using backward-chaining reasoning.
    ***COPY ELEMENT.KBS;1 INPUT.DAT;3

PLEASE INPUT 1 TO RUN THIS PROGRAM?
1

$ RUN INFER

DID YOU ASSIGN THE KNOWLEDGE BASE FILE AS "INPUT" FILE? OR COPY TO "INPUT.DAT" FILE?
1 → YES, 0 → NO
1
    READING RULES

DID YOU WISH TOI USE FACTS IN "FACTS.DAT" FILE?
1 → YES, 0 → NO
1PLEASE CHOOSE INFERENCE PROCESS!
1 → FORWARD CHAINING (YOU ALREADY KNOW SOME FACTS!)
2 → BACKWARD CHAINING
3 → SPECIAL REASONING (ASK QUESTIONS FOR EVERY RULE!)
2
    BACKWARD CHAINING PROCESS

PLEASE ANSWER THE FOLLOWING QUESTION WITH A FACTOR RANGING FROM 0.0 (DEFINITE "NO") TO 1.0 (DEFINITE "YES")
??
This problem is a mechanism problem.
0.0

PLEASE ANSWER THE FOLLOWING QUESTION WITH A FACTOR RANGING FROM 0.0 (DEFINITE "NO") TO 1.0 (DEFINITE "YES")

??

This problem is beam or frame.
1.0

PLEASE ANSER THE FOLLOWING QUESTION WITH A FACTOR RANGING FROM 0.0 (DEFINITE "NO" TO 1.0 (DEFINITE "YES")
??

Only one plane of bending or axial force for this problem.
0.5

PLEASE ANSWER THE FOLLOWING QUESTION WITH A FACTOR RANGING FROM 0.0 (DEFINITE "NO") TO 1.0 (DEFINITE "YES")
??

This problem has applied torsional force.
0.0

PLEASE ANSWER THE FOLLOWING QUESTION WITH A FACTOR RANGING FROM 0.0 (DEFINITE "NO") TO 1.0 (DEFINITE "YES")
??

This problem is truss structure.

PLEASE ANSWER THE FOLLOWING QUESTION WITH A FACTOR RANGING FROM 0.0 (DEFINITE "NO") TO 1.0 (DEFINITE "YES")
??

Only extensional vibration for this problem.
0.2

PLEASE ANSWER THE FOLLOWING QUESTION WITH A FACTOR RANGING FROM 0.0 (DEFINITE "NO") TO 1.0 (DEFINITE "YES")
??

Only axial force for this problem.
0.0

PLEASE ANSWER THE FOLLOWING QUESTION WITH A FACTOR RANGING FROM 0.0 (DEFINITE "NO") TO 1.0 (DEFINITE "YES")
??

This problem has more than one plane of bending and any combination of axial, bending and torsional loads.
0.9.

****

THE HYPOTHESIS THAT HAS CERTAINTY 0.900

This problem requires elements E21. The E21 are general straight or circularly curved beam elements, such as channels, wide-flanges, angles, tubes, zees.

OPTIONS ARE
1 → SEE AN EXPLANATION
2 → CONTINUE TO NEXT CASE
3 → EXIT
? PLEASE ANSWER WITH THE NUMBER !
1

THIS IS RULE 8

Ch. 10] ANALYSIS AND DESIGN OF STRUCTURE SYSTEMS 205

YOU HAVE 1.00 CERTAINTY FOR →
This problem has more than one plane of bending and any combination of axial, bending and torsional loads.

SO THE FOLLOWING HYPOTHESIS HAS 0.900 CERTAINTY →
This problem requires elements E21. The E21 are general straight or circularly curved beam elements, such as channels, wide-flanges, angles, tubes, zees.
⋮

$ RUN INFER
⋮

PLEASE ANSWER THE FOLLOWING QUESTION WITH A FACTOR RANGING FROM 0.0 (DEFINITE "NO") TO 1.0 (DEFINITE "YES")
??
The FEM elements have been selected for this problem.
1.0

PLEASE ANSWER THE FOLLOWING QUESTION WITH A FACTOR RANGING FROM 0.0 (DEFINITE "NO") TO 1.0 (DEFINITE "YES")
??
This structure is modeled by one-dimensional line elements.
0.9

****
THE HYPOTHESIS THAT HAS CERTAINTY 0.900

You should run the MESH1 knowledge base to generate mesh by using backward-chaining reasoning.
***COPY MESH1.KBS;1 INPUT.DAT;3

PLEASE INPUT 1 TO RUN THIS PROGRAM?
1

$ RUN INFER
⋮

PLEASE ANSWER THE FOLLOWING QUESTION WITH A FACTOR RANGING FROM 0.0 (DEFINITE "NO") TO 1.0 (DEFINITE "YES")
PLEASE ANSWER THE FOLLOWING QUESTION WITH A FACTOR RANGING FROM 0.0 (DEFINITE "NO") TO 1.0 (DEFINITE "YES")
??
This problem is only under static loading.
1.0

PLEASE ANSWER THE FOLLOWING QUESTION WITH A FACTOR RANGING FROM 0.0 (DEFINITE "NO") TO 1.0 (DEFINITE "YES")
??
This problem only considers yielding case.
0.95

PLEASE ANSWER THE FOLLOWING QUESTION WITH A FACTOR RANGING FROM 0.0 (DEFINITE "NO") TO 1.0 (DEFINITE "YES")

206 DESIGN OPTIMIZATION   [Pt. II

??
This problem is only under concentrated loading.
0.1

PLEASE ANSWER THE FOLLOWING QUESTION WITH A FACTOR
RANGING FROM 0.0 (DEFINITE "NO") TO 1.0 (DEFINITE "YES")
??
This problem has distributed loading.
1.0

PLEASE ANSWER THE FOLLOWING QUESTION WITH A FACTOR
RANGING FROM 0.0 (DEFINITE "NO") TO 1.0 (DEFINITE "YES")
??
This problem has axes of symmetry for both the geometry and loading.
0.0

PLEASE ANSWER THE FOLLOWING QUESTION WITH A FACTOR
RANGING FROM 0.0 (DEFINITE "NO") TO 1.0 (DEFINITE "YES")
??
This problem is unsymmetric in geometry or loading.
1.0

PLEASE ANSWER THE FOLLOWING QUESTION WITH A FACTOR
RANGING FROM 0.0 (DEFINITE "NO") TO 1.0 (DEFINITE "YES")
??
This problem is a beam problem.
0.5

PLEASE ANSWER THE FOLLOWING QUESTION WITH A FACTOR
RANGING FROM 0.0 (DEFINITE "NO") TO 1.0 (DEFINITE "YES")
??
This problem only needs displacement information.
0.5

PLEASE ANSWER THE FOLLOWING QUESTION WITH A FACTOR
RANGING FROM 0.0 (DEFINITE "NO") TO 1.0 (DEFINITE "YES")
??
This problem needs stress informations.
1.0

PLEASE ANSWER THE FOLLOWING QUESTION WITH A FACTOR
RANGING FROM 0.0 (DEFINITE "NO") TO 1.0 (DEFINITE "YES")
??

This problem is frame problem.
1.0

****
THE HYPOTHESIS THAT HAS CERTAINTY 0.950

Choose nodes at following places: support positions, joint locations, at each end
of the frame member, at concentrated loads position and points where the

distributed loads change significantly, also choose nodes where response may be required. If distributed loads range is long, choose some nodes in the range of the distributed loads. Also choose 5 to 10 elements for each frame member and try to make the element lengths equal.

Example 3: design modelling for cantilevered beam with fixed tip mass.
$ RUN INFER
⋮

??
This problem deals with vibration analysis or is subjected to dynamic loading.
1.0
PLEASE ANSWER THE FOLLOWING QUESTION WITH A PROBABILITY FACTOR RANGING FROM 0.0 (LEAST CERTAIN) TO 1.0 (MOST CERTAIN)
??
The nonstructural mass in this problem is four to five times larger than the structural mass.
1.0
PLEASE ANSWER THE FOLLOWING QUESTION WITH A PROBABILITY FACTOR RANGING FROM 0.0 (LEAST CERTAIN) TO 1.0 (MOST CERTAIN)
??
This problem is statically determined problem.
1.0
⋮

??
The geometry of the structure is fixed.
1.0
⋮

??
This problem is a beam or frame problem.
1.0
****
THE HYPOTHESIS THAT HAS CERTAINTY 1.0
Should choose the cross-sectional properties (A, I, J) as design variables. This would improve the linear assumption for frequency constraints.
⋮

****
THE HYPOTHESIS THAT HAS CERTAINTY 1.0
Select natural vibration fequencies to formulate inequality constraints. The frequency can be larger than lower bounds or less than upper bounds. The constraint is directly written in terms of the squares of the circular frequency.

11

A knowledge-based framework for constraint activity identification in optimal design of aircraft structures

P. Y. Papalambros
The University of Michigan

1. INTRODUCTION

In the final stages of the design process, when a configuration has been essentially achieved and a refinement of proportions is sought, design decisions may be represented by a precise mathematical optimization statement, expressed as the non-linear programming (NLP) model:

$$\begin{aligned}
\text{minimize} \quad & f(\mathbf{x}) && \mathbf{x} \in \mathcal{X} \\
\text{subject to} \quad & h_j(\mathbf{x}) = 0 && j = 1, 2, \ldots, m_1 \\
& g_j(\mathbf{x}) \leq 0 && j = m_1 + 1, m_1 + 2, \ldots, m_1 + m_2
\end{aligned} \quad (1)$$

Here f, h and g are the objective, equality and inequality constraints respectively, $\mathbf{x}$ is the vector of design variables $x_1, x_2, \ldots, x_n$, and the set $\mathcal{X}$ is a subset of $\mathcal{R}^n$, included in the statement to indicate that there may exist other restrictions on the variables that are not explicit in the constraints, for example, integer values. The objective is shown as a single objective, since multiobjective formulations are usually transformed to a single objective. The above mathematical statement is referred to as the **design optimization model**. A thorough introduction to the formulation and solution of such models may be found in [18].

The design model represented in eqn. (1) implicitly assumes the existence of an analysis model used to evaluate the functions f, h, g. For example, analysis models may be finite element analyses of structural behaviour, or system simulations. These models may be explicit or implicit. In an explicit model, the functions are represented by direct evaluation of mathematical, usually algebraic, expressions; in an implicit model, the functions are typically represented by a numerical procedure, for example, iterative solution of a set of differential equations. Complex structural models are implicit, and the evaluation of model functions has a high computational cost.

From a mathematical viewpoint, explicit models are preferable since optimality asssertions about them are easier, for example, by detecting mathematical properties such as monotonicity or convexity; also, it is often possible to manipulate the model in order to extract additional information or properties. From a practical viewpoint, the models in many important classes of design optimization problems are implicit, posing a substantially increased burden on ascertaining the reliability of the optimization results. However, in practice one seldom seeks a mathematical optimum, aiming only at satisfying designs [21]; thus the disadvantage of implicit models is significantly tempered. In certain cases, as in the presence of monotonicity in the model, qualitative results may be obtained even when the models are implicit [18].

From an operational viewpoint, it is useful to distinguish two phases in the optimization process. When the optimal design model has been precisely formulated, the modelling phase of the optimization procedure concludes, and is followed by the solution phase. For most realistic models, the solution phase will include an iterative procedure that should converge to the optimum. An iteration begins by selecting an initial design corresponding to a starting value for x and executing the calculation represented by the analysis model(s), so the values for objective and constraint functions are obtained. The cost of subsequent calculations for evaluating the next iterant design increases with the size of the problem, as measured by the number of variables and constraints.

In structural design models, the number of constraints is quite large relative to the number of independent variables. In addition, the computational cost for calculating constraint functions and their gradients is the overriding concern during a structural optimization study. Thus the single most important overall efficiency criterion is the number of constraint evaluations. This criterion motivates the present article.

The key strategy for dealing with a large number of constraints is to select only a subset of the original constraint set at any given iteration. If the solution to an optimization model changes when a constraint is removed from the model, then this constraint is called active. An active constraint is a direct contributor to optimality and corresponds either to an equality constraint (such as equilibrium) or to a critical design requirement (such as setting the maximum stress equal to the strength of the material used). When a subset of the original constraints is selected for further processing, these constraints together with the objective function constitute a submodel of the original problem. The submodel contains only those constraints that are known to be active or that have been judiciously selected as having a high likelihood to become active. A set of rules that is used to generate the submodel is called a submodel formulation strategy or an active set strategy. Thus we have the following definitions.

> An **active set strategy** is a set of rules used to decide which constraints may be active at the optimum. An **active set** $\mathcal{A}$ is a set that contains only those constraints that are known to be definitely active at the current iteration, meaning that they are satisfied as

equalities. An **extended active set** $\mathscr{E}$ is a set that contains both the currently active constraints and additional constraints that are considered candidates for activity in one or more subsequent iterations.

(For exceptions of active constraints that are not satisfied as equalities, and for equalities that may not be active, as well as other definitions related to constraint activity, see [18]. Clearly, at any given point the active set is a subset of the extended set.

The rules in an active set strategy will apply in general to both an active set and an extended set. In the mathematical programming literature the term 'active set strategy' applies only to the active set, while in the structural optimization literature the term usually applies to the extended active set. Here the term is used to cover both cases, a use that becomes natural when a knowledge-based approach is followed.

The discussion and concepts involved are general, but the motivation and specific implementation example come from aircraft structural design and the optimization program OPTSYS [3,4,17].

2. GENERALIZED QUALITATIVE MODELS

We can recognize three forms of knowledge applicable to the design optimization process:

Global knowledge pertains to facts independent of a particular design point and equally true at all points in the design space. **Local knowledge** pertains to facts applicable to a particular design point. **Evolutionary knowledge** pertains to facts observed or deduced from local knowledge about a sequence of design points generated by a particular algorithm.

This classification allows a more precise description of the source of knowledge utilized. Global knowledge is a result of analysis, experience, or intuition. Local knowledge is generated through numerical computation. All customary non-linear optimization codes operate mostly on such knowledge. Evolutionary knowledge is a mixture of both computation and judgement. These definitions are useful in an operational way; that is, they refer to knowledge used explicitly during the process. Otherwise, one may argue that the construction of an algorithm contains itself global knowledge and the distinction becomes blurred. However, if we restrict the definition to knowledge explicitly utilized at decision points during optimization, then the classification would be a useful methodological tool.

Sources of knowledge are as follows: designer's experience or intuition, analyst's experience or intuition, databases of previous results, mathematical proofs, and computation. Note that the distinction between designer and analyst is a practical one, as in most organizations today these are two different individuals (usually humans, but also computer programs). For simplicity we may refer to the designer or analyst as the same source of knowledge. Databases of previous results may not be always formal, but may simply reside in human or computer memory, or in paper files. In order to utilize either global or evolutionary knowledge, it is important that these data bases are formalized in a way that data may be organized, reduced, and retrieved when making decisions. Mathematical proofs and local computation are the usual tools that have been employed to date.

In order to utilize different forms of knowledge explicitly, this knowledge must be properly codified. The developments in artificial intelligence methods make it increasingly easier to represent and manipulate these forms of knowledge in a practical way. The connection between AI and design optimization was first made by Azarm and Papalambros [2], including use of symbolic algebra processors [10]. Since then much research has aimed at the development of knowledge-based systems for optimal design; see for example [1,7,15] for reviews.

We now proceed to describe how implicit models may be used qualitatively to derive explicit information about the model. This is done in the context of the knowledge-based system PRIMA (production system-based implicit elimination in monotonicity analysis) developed for this purpose [19].

PRIMA is a rule-based system implemented in the OPS5 production system development tool [6,8], and deals primarily with generating global knowledge about constraint activity (Fig. 11.1). The main tool for this is the

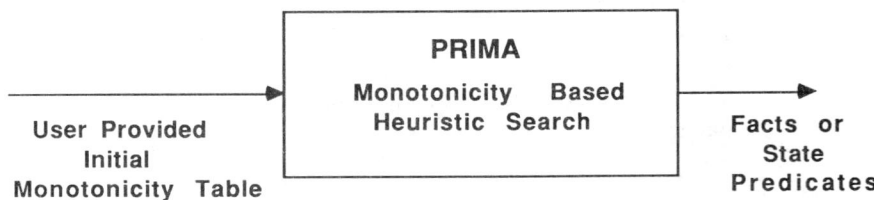

Fig. 11.1 — Outline of the PRIMA system.

monotonicity table of the NLP model [18,20] and the implicit elimination transformation of monotonic models developed in [14]. The monotonicity table is simply a table with columns and rows associated with variables and model functions respectively. The table entries are the monotonicities of the function with respect to each of the variables (indicated by '+' for increasing, '−' for decreasing, 'U' for unknown or non-monotonic and an empty entry if the variable does not appear in the corresponding function). This allows a rapid visual application of boundedness arguments using the monotonicity principles; for further details, see [18]. This table can serve as the basic knowledge construct for a production system that identifies constraint activity properties.

A set of heuristic rules for applying implicit elimination was presented in [16] and is the basis for the PRIMA implementation. The knowledge flow diagram for PRIMA is given in Fig. 11.2. The intention of developing the

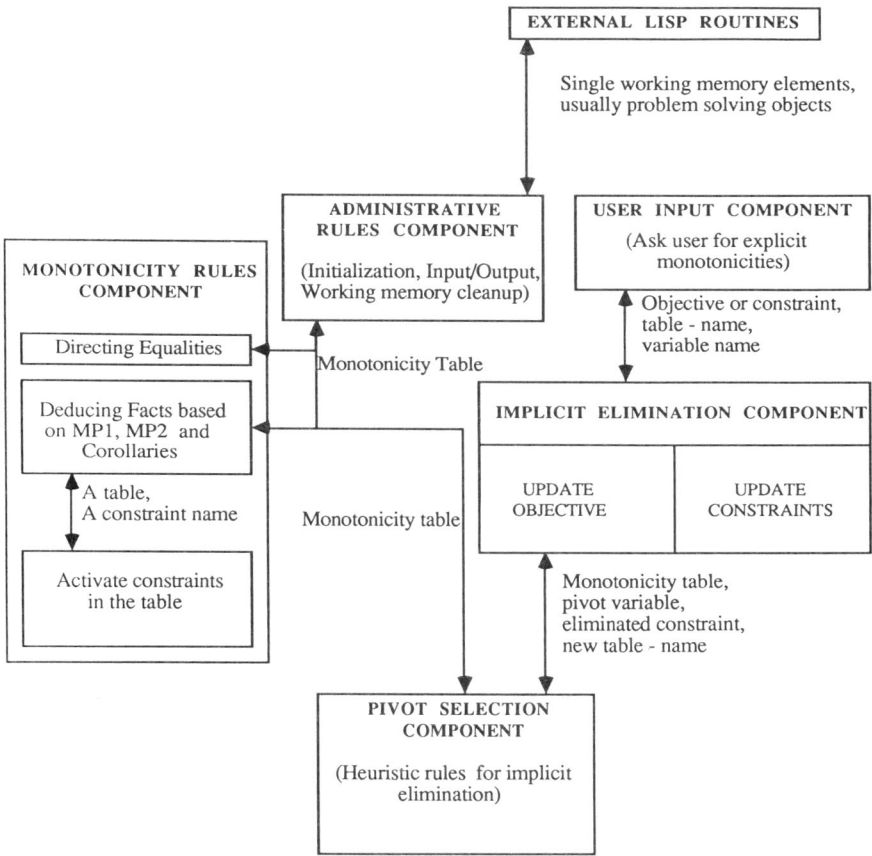

Fig. 11.2 — Knowledge flow diagram for PRIMA.

system was to generate automatically global information about constraint activity and model boundedness which can then be given directly to a global or local–global active set strategy [9, 11–13]. This removes, at least in part, the difficulty that relatively unskilled users encounter in discovering rigorous global knowledge; it eliminates tedious procedures prone to errors, when manually performed; and it can be applied to implicit models, if sufficient monotonicity knowledge is available. The net result is usually either identification of necessary activity properties, or determination of a poorly bounded model.

The monotonicity principles [18] and several rules in the form of corollaries, together with some heuristics, are at the core of the knowledge base coded in PRIMA. The monotonicity table forms the basic abstract data structure in the problem domain. The rules are applied to obtain facts about the design model. These facts, called **state predicates**, are derived from necessary conditions and must be satisfied by every optimum solution, called a **state**. Ideally, these facts are derived one by one and the system automatically generates LISP-like macros in the form of predicate functions, each of which should evaluate to true at every optimal design or state.

Constraints identified as active (equalities and inequalities) may be eliminated, but usually this procedure cannot be effected explicitly. Implicit elimination of active constraints reduces the problem size and results in a new derived monotonicity table which can generate more state predicates. This process may possibly continue until all active constraints are eliminated. However, a complete search of the implicit elimination tree is neither practical nor desired. In fact, one needs to be extremely selective in expanding the nodes during the search process, as indicated by the heuristic rules in [16]. Implicit elimination can be performed in a systematic manner and special advantage can be derived often by directing an equality first [18]. In theory, all active equality constraints can be directed such that the optimum solution of the relaxed model lies in the infeasible region.

Implicit elimination using just the monotonicity information can lead to unknown monotonicities in the resulting expressions. If the search seems promising along this path, explicit elimination using symbolic mathematics may be of use. When further following of a particular path is deemed unprofitable, backtracking is performed. Full or partial traversal of this implicit elimination tree yields state predicates embodying global information. Although the obtained knowledge is generally incomplete, this does not indicate an unsuccessful process, because the intention is not to solve the problem entirely by global methods but to use whatever facts are generated here in a judicious global–local strategy. To summarize, the primary task is to search the implicit elimination tree. Visiting a node, i.e. monotonicity table, consists of applying mathematically rigorous necessary conditions implemented in the form of rules; the tree arcs (connecting one node to another) correspond to the process of implicit elimination; and the choice of pivots used in elimination (corresponding to selective expansion of nodes) is based on heuristic rules. This classification of rigorous rules (at the nodes) and hueristic rules (at the arcs) can be considered a characteristic of this problem domain. Further details on PRIMA may be found in the cited reference.

The main point in this section is that the monotonicity table is a generalized model on the optimal design problem, and constraint activity results may be derived from it. Obviously, this model is not always available. In certain cases, an approximate model may serve for constructing the monotonicity table and deriving global knowledge from it. Results obtained from the complete model should corroborate this global knowledge; if not, the source of inconsistency should be sought either in the approximation or in the numerical (local) determination of optimality.

3. GENERALIZED ACTIVE SET STRATEGIES

Assume now that the structural optimization model is in the negative null form of eqn (1) and that there are no equality constraints present explicitly in the optimization model. In most optimization algorithms, at each iteration, a point x_k is used to generate a new point x_{k+1} which is then added to the sequence of design points $\{x_k\}$ that should converge to the optimum. To test for termination or to prepare for the new iteration, at least the values of constraints and gradients must be recalculated at point x_{k+1}. In this section, a general strategy for handling constraint activity is presented. A summary of symbols and terminology is given in Table 11.1.

At a given iteration point x_k all constraints are assigned a priority index $p_{i,k}$ corresponding to constraint g_i, defined as follows:

$$p_{i,k} = (g_i/g_{i,\text{lim}})_k \tag{2}$$

where $g_{i,\text{lim}}$ is the upper bound (limit) for the constraint; that is, all constraints are assumed to be in the form

$$g_i \leq g_{i,\text{lim}} \tag{3}$$

To simplify notation, the subscript k is omitted in the subsequent discussion. Special mention for the initial iteration will be made later.

The true active set would contain only those constraints that have $p_{i,k} = 1$, since these will be satisfied as strict equalities. However, in the extended active set we include also constraints that are likely to be active at the optimum. In many cases, it is only this likelihood that we can express, particularly at iterations away from the optimum. Thus the priority index should contain not only a measure of current infeasibility but also a measure of likelihood, or bias, for future activity. We can then define a biased priority index, $P_{i,k}$, as follows;

$$P_{i,k} = p_{i,k} + B_{i,k} \tag{4}$$

where $B_{i,k}$ is a measure of bias introduced by the rules controlling decisions in the active set strategy. More specifically, if a constraint belongs to the active set $\mathcal{A}$, then it should be also included in $\mathcal{E}$. This means that a large value will be assigned to the bias. Formally,

$$\text{if} \quad p_{i,k} \leq 1 + \varepsilon_f \quad \text{and} \quad \mu_{i,k} > \varepsilon_l, \quad \text{then} \quad B_{i,k} = C_{1,i}, \tag{5}$$

where ε_f is the feasibility tolerance, ε_l is the tolerance for positivity of the Lagrange multiplier estimates $\mu_{i,k}$, and $C_{1,i}$ is a large number. The total number of constraints satisfying eqn. (5) at a given iteration is represented by the symbol $M_{\text{AS}}(x_k)$.

Furthermore, if a constraint violates its bound, it may qualify for inclusion in $\mathcal{E}$. Formally

$$\text{if} \quad p_{i,k} \geq \varepsilon_v, \quad \text{then} \quad B_{i,k} = C_{2,i} \tag{6}$$

where ε_v is a bound violation tolerance not directly related to ε_f, and $C_{2,i}$ is a bias that may have any value, provided that $C_{2,i} < C_{1,i}$. The total number of

Table 11.1 — Nomenclature for parameters in a generalized active set strategy

$B_{i,k}$	bias for constraint i at iteration k
C_1	parameter for bias associated with constraints in $\mathcal{A}'(\mathbf{x}_k)$
C_2	parameter for bias associated with constraints in $\mathcal{V}(\mathbf{x}_k)$
C_3	parameter for multiple of N as upper bound on size of $\mathcal{E}$
C_4	parameter for fraction of $M_{\mathrm{ES,max}}$ in extra constraints
C_5	parameter for small fixed number of extra constraints
C_6	parameter for bias associated with constraints in $\mathcal{F}(\mathbf{x}_k)$
C_7	parameter for minimal inclusion of frequency constraints
C_8	parameter for lower limit on flutter constraints
C_9	parameter for bias associated with constraints in $\mathcal{L}(\mathbf{x}_k)$
C_{10}	parameter for minimal inclusion of frequency constraints
ε_b	tolerance for feasible bound proximity
ε_{bf}	tolerance for frequency constraints
ε_{bl}	tolerance for flutter constraints
ε_f	tolerance for feasibility
ε_l	tolerance for positivity of multipliers
ε_v	tolerance for inclusion to set $\mathcal{V}$
$f_{i,k} = (g_i/g_{i,\lim})$	for frequency and flutter constraints
$g_{i,\lim}$	upper bound on constraint function g_i
L	cutoff number of constraints for considering a model as not 'large'
M	total number of constraints in the model
$M_{\mathrm{AS}}(\mathbf{x}_k)$	number of stress constraints in $\mathcal{A}'$ with positive multipliers at $\mathbf{x}_k$ (similarly for other constraint types)
$M_{\mathrm{BS}}(\mathbf{x}_k)$	number of feasible stress constraints near their bounds at $\mathbf{x}_k$ (similarly for other constraint types)
$M_{\mathrm{VS}}(\mathbf{x}_k)$	number of stress constraints violated at $\mathbf{x}_k$ (similarly for other constraint types)
M_{BVF}	total number of constraints in $\mathcal{F}$
M_{BVL}	total number of constraints in $\mathcal{L}$
M_{E}	total number of constraints in $\mathcal{E}$
M_{EA}	total number of aeroelastic constraints in $\mathcal{E}$
M_{EB}	total number of buckling constraints in the $\mathcal{E}$
M_{ED}	total number of deflection constraints in $\mathcal{E}$
M_{EF}	total number of frequency constraints in $\mathcal{E}$
M_{EL}	total number of flutter constraints in $\mathcal{E}$
M_{EO}	total number of other constraints in $\mathcal{E}$
M_{ES}	total number of stress constraints in $\mathcal{E}$
$M_{\mathrm{ES,max}}$	upper bound on number of stress constraints in $\mathcal{E}$
M_{A}	total number of aeroelastic constraints in the model
M_{B}	total number of buckling constraints in the model
M_{D}	total number of deflection constraints in the model
M_{F}	total number of frequency constraints in the model
M_{L}	total number of flutter constraints in the model
M_{O}	total number of other constraints in the model
M_{S}	total number of stress constraints in the model
N	number of design variables
$P_{i,k}$	biased priority index for inclusion in $\mathcal{E}$
$p_{i,k}$	priority index for inclusion in $\mathcal{E}$

constraints that satisfy eqn. (6) at a given iteration is represented by the symbol $M_{BV}(\mathbf{x}_k)$. All such constraints are said to belong in a set $\mathcal{V}$.

After the priority list has been generated, the number of constraints to be included in $\mathcal{E}$ must be decided. The selection of this number is one of the most critical decisions affecting computational cost and number of iterations. The computational cost per iteration is associated with calculating the gradients for all members of $\mathcal{E}$, so their number should be kept as small as possible. On the other hand, the optimizer will solve the subproblem containing only these selected constraints and identify violations of the excluded constraints after the iteration has been completed. Attempting to recover from infeasibility usually leads to increased number of iterations. Thus from this viewpoint the number of constraints in $\mathcal{E}$ should be large enough to contain all likely active constraints so that the overall cost of all iterations may be kept as low as possible. The number of constraints to be included in $\mathcal{E}$ must be chosen so as to balance these two competing requirements.

A general procedure for selecting this number, suitable for knowledge-based implementation, is given below. This procedure is obviously necessary when the model has a large number of constraints, otherwise all constraints may be included in each iteration. What constitutes a 'large' number is a subjective judgement based on the amount of resources available to solve the problem, and the space and other limitations of the analysis modules that are used.

The total number of constraints in the extended active set, M_E, is the sum of the numbers of constraints of different types in that set. In an aircraft structural design we set

$$M_E = M_{ES} + M_{ED} + M_{EB} + M_{EA} + M_{EF} + M_{EL} + M_{EO} \qquad (7)$$

Here the first subscript E stands for an expected activity (hence inclusion in the extended active set) and the second subscript S stands for stress constraints, D for deflections, B for buckling, A for aeroelastic constraints, F for frequency, L for flutter, and O for other constraints that may be included in the model. Note that for practical purposes the buckling constraints may be taken as a special case of the stress constraints. These numbers are less than or equal to the corresponding numbers, M, M_I, in the complete model, namely

$$M_E \leq M \quad \text{and} \quad M_{EI} \leq M_I \quad \text{for} \quad I = S, D, B, A, F, L, O \qquad (8)$$

The largest number is usually M_S and so selection of M_{ES} is the major part of selecting M_E.

3.1 Stress constraints

Let us now examine how this number of constraints, M_{ES}, is determined. We will assume the example of stress constraints in aircraft structural design that employs composite materials, so the underlying analysis model is a (large) finite element model.

After the original model has been formulated, one may count explicitly the number of stress constraints that have been included, thus obtaining M_S. In a typical problem this number may be estimated as given by the product

$$M_S = (\text{number of elements}) \times (\text{number of layers}) \times \text{number of stress points} \times (\text{number of loading cases}) \times (\text{number of failure criteria}) \quad (9)$$

The term 'number of failure criteria' is used in a somewhat generalized way. The failure criteria may be maximum normal stresses, maximum shear stress, von Mises equivalent stress, or maximum strain, i.e.

$$\begin{array}{llll} \sigma_x \leq \sigma_{x,\lim} & \sigma_y \leq \sigma_{y,\lim} & \sigma_z \leq \sigma_{z,\lim} & \\ \tau_{xy} \leq \tau_{xy,\lim} & \sigma_{vM} \leq \sigma_{vM,\lim} & \varepsilon_x \leq \varepsilon_{x,\lim} & \end{array} \quad (10)$$

but also local buckling. The number M_S is large when no knowledge is used to decide an *a priori* elimination of constraints that are very unlikely to be active. For example, from the failure criteria above usually only one or two can be active. If the total number of variables is N, the maximum possible number of active constraints cannot exceed N, so

$$M_{AS} \leq N \quad (11)$$

where the first subscript A is used to indicate a constraint that is actually active at the optimum, as opposed to expected activity indicated by the first subscript E. It is then reasonable to assume that the number of expected active constraints should be at least N but not much larger than N, namely

$$N \leq M_{ES} \leq C_3 N \quad (12)$$

where C_3 is a multiplication factor usually greater than unity. An upper bond on M_{ES} is set by

$$M_{ES,\max} = \max\{M_S, L\}, \quad (13)$$

where L is the cut-off parameter for using the active set strategy.

A more satisfactory way for setting this upper limit is as follows:

$$M_{ES,\max} = \max\{\min[M_S, L], C_3 N\} \quad (14)$$

This way, for a large number of constraints the control will be provided by how many variables exist in the model plus possible heuristic rules for determining a value for C_3. For a small number of constraints, all of them will be included. In the usual case where $N \geq M_S$, we may select

$$C_3 = M_S/N \quad (\text{for } N \geq M_S) \quad (15)$$

The number $M_{ES,\max}$ selected in the manner just discussed is only an upper bound on the desired number M_{ES}. This number may now be selected according to the expression

$$M_{ES} = \min\{M_{AS}(\mathbf{x}_k) + \max[M_{VS}(\mathbf{x}_k), C_4 M_{ES,\max} + C_5], M_{ES,\max}\} \quad (16)$$

where $M_{AS}(\mathbf{x}_k)$ and $M_{VS}(\mathbf{x}_k)$ are the numbers of constraints that, at $\mathbf{x}_k$, had positive multipliers and were violating their bounds respectively, as given by eqns (5) and (6). The parameter C_4 has values less than unity and the parameter C_5 has a small integer value; these two parameters together enforce the requirement that a fraction of the maximum allowable number plus an additional few constraints are included as 'extra', to ensure that enough constraints will be included in $\mathscr{E}$ should the numbers $M_{AS}(\mathbf{x}_k)$ and $M_{VS}(\mathbf{x}_k)$ be too small. Finally, special treatment is needed for the initial iteration, $k=1$. Here there is more limited information, so a simple decision is to set

$$M_{ES} = M_{ES,\max} \tag{17}$$

This is usually sufficient for initializing the algorithm.

3.2 Deflection constraints

The strategy for deflection constraints is the same as for stresses and need not be repeated. The only differences exist in the selection of some program parameters. In a typical problem the total number of deflection constraints is

$$M_D = \text{(number of user-defined deflection constraints)} \times \text{(number of loading cases)}. \tag{18}$$

This number is relatively small so we may set

$$M_{ED,\max} = \max\{\min[M_D, L], C_3 N\} \approx M_D$$

$$M_{ED} = \min\{M_{AD}(\mathbf{x}_k) + \max[M_{VD}(\mathbf{x}_k), C_4 M_{ED,\max} + C_5], M_{ED,\max}\} \tag{19}$$

3.3 Frequency constraints

The case of frequency constraints is somewhat different. The functions calculated by the analysis model need a lower bound:

$$g_i \geq g_{i,\lim} \tag{20}$$

So the priority index should be defined by

$$p_{i,k} = 1/f_{i,k} \tag{21}$$

where

$$f_{j,k} = (g_i/g_{i,\lim})_k \tag{22}$$

In general, frequency constraints should provide both upper and lower bounds, so that the design lies in an acceptable range away from resonance. Upper bounds can be treated as those of earlier constraint types. Then the active set strategy can be the same as for stress constraints.

The number of frequency constraints to be selected for the active set, M_{EF}, is the same as the number of eigenvalues that must be calculated by the analysis model. The maximum number specified, $M_{EF,\max}$, sets the corresponding parameters in the analysis model, so it should be kept low. Unless other knowledge is available, it is not clear how many eigenvalues should be selected. Therefore, a simple criterion suggested for the current frequency constraint types is as follows:

$$\text{if} \quad f_{i,k} < \varepsilon_{bf}, \quad \text{then} \quad B_{i,k} = C_{6,i} \tag{23}$$

Here ε_{bf} is a parameter with a value greater than 1, typically equal to 2 or 3. It is used to identify all constraints that are either violated or not sufficiently away from the bound. The set containing all such constraints, $\mathscr{F}$, has cardinality M_{BVF} and contains all the constraints of which some will be selected for inclusion in $\mathscr{E}$. The parameter $C_{6,i}$ is only significant in enforcing that some frequency constraints will be included in $\mathscr{E}$, rather than substantially distinguishing among the frequency constraints themselves.

The number of constraints that are included in $\mathscr{E}$ may be selected as

$$M_{EF} = \max\{C_7, min[M_{BVF}(\mathbf{x}_k), M_{EF,max}]\} \tag{24}$$

Here C_7 is a parameter that guarantees the inclusion of some frequency constraints, if they are present at all; the number $M_{EF,max}$ is defined by the user. One difficulty with this approach is that, effectively, the number $M_{BVF}(\mathbf{x}_k)$ is not utilized in the active set strategy since the value of $M_{EF,max}$ dominates the decision in eqn. (24). In current practice the value of $M_{EF,max}$ is set at the beginning of the iterations, and is only changed (manually) if it is detected, by the iteration history, that $M_{EF,max}$ is so small that significant constraints from the set $\mathscr{F}$ are excluded. An alternative is to set

$$M_{EF} = \max\{M_{BVF}(\mathbf{x}_k), M_{EF,max}\} \tag{25}$$

and to link the analysis software with the optimizer so that only calculations for M_{EF} frequencies are performed per iteration.

As noted further below, from a knowledge-based implementation viewpoint, frequency constraints should be treated just like stress constraints, with proper identification of current activity, infeasibility, or proximity to the bound. Rules about selecting the various parameters can contain any desirable special considerations.

3.4 Flutter constraints

Flutter constraints also are in a lower bound form like the frequency constraints, i.e. a lower bound is placed on the damping of a flutter mode [5]:

$$g_i \geq g_{i,lim} = C_8 \tag{26}$$

where C_8 is a user-defined parameter. The priority index should be again defined by

$$p_{i,k} = 1/f_{i,k} \tag{27}$$

where

$$f_{i,k} = (g_i/g_{i,lim})_k \tag{28}$$

All flutter constraints are ranked according to a biased priority index as usual. Following the description for the frequency constraints we have

$$\text{if} \quad f_{i,k} < \varepsilon_{bl}, \quad \text{then} \quad B_{i,k} = C_{9,i} \tag{29}$$

Here ε_{bl} is a parameter used to identify all constraints that are not very far from the bound. The set containing all such constraints, $\mathscr{L}$, has cardinality M_{BVL} and contains all the constraints of which some will be selected for inclusion in $\mathscr{E}$. The parameter $C_{9,i}$ is only significant in enforcing that some flutter constraints will be included in $\mathscr{E}$, rather than substantially distinguishing among the flutter constraints themselves.

The number of flutter constraints is generally given by

$$M_L = \text{(number of degrees of freedom in the analysis model)} \times \text{(number of flight cases)} \quad (30)$$

where the number of flight cases is given by the number of Mach numbers multiplied by the number of altitudes. The number of constraints that are included in $\mathscr{E}$ may be selected as

$$M_{EL} = \max\{C_{10}, \min[M_{BVL}(\mathbf{x}_k), M_{EL,\max}]\} \quad (31)$$

Here C_{10} is a parameter that guarantees the inclusion of some flutter constraints, if they are present at all; the number $M_{EL,\max}$ is defined by the user.

3.5 Knowledge-based implementation

The values for $C_{1,i}$ and $C_{2,i}$ should discriminate among the different constraints. In current practice they are fixed, and the different constraint types (i.e. stress, deflection, etc.) are treated completely independently. However, the priority numbers should depend on the constraint they are applied to. First, constraints can be classified as physical and practical [18]. If the constraint is practical, it should not be active unless the problem is otherwise unbounded. Therefore a low priority value $C_{1,i}$ may be assigned to such a constraint at least in the earlier iterations, attempting to keep it outside the active set. Note that this must be done by assigning values of $C_{1,i}$ and $C_{2,i}$ for practical constraints.

Next, physical constraints are characterized by the actual specification they represent, i.e. stress or deflection limits, buckling, and so on. This characterization must be used in assigning values to $C_{1,i}$, as there may be preferences expressed by the designer–analyst based on qualitative or other knowledge that has not been used explicitly elsewhere. Usually, constraints other than stress may not be large in number compared with the stress constraints, so this subjective input may not be often of practical use. However, the same argument may be applied to different types of stress constraints. Stress constraints tend to apply locally, at an element or a component level, at several points near each other. Quite often it is possible to express some local dominance knowledge about these constraints and to select one or two as the most likely to be critical. This may not be so important in the selection of $C_{1,i}$, unless there is suspected degeneracy, but it may be used in the selection of $C_{2,i}$ values.

Another issue is related to assuming that constraints belong to the active set $\mathscr{A}$ simply based on positivity of the multiplier values. In convex approximation methods, favoured in structural optimization, the dual solution is utilized and the values of the Lagrange multipliers at a given iteration are estimated directly as the solution of the problem that is dual to the convex approximation of the submodel. The first-order optimality conditions are the Karush–Kuhn–Tucker (KKT) conditions

$$\partial f/\partial x_i + \sum_j \mu_j(\partial g_j/\partial x_i) = 0 \text{ for all variables } i \quad \text{(stationarity)}$$

$$g_j \leq 0 \text{ for all constraints } j \quad \text{(feasibility)} \quad (32)$$

$$\mu_j g_j = 0 \quad \text{(transversality)}$$

$$\mu_j \geq 0 \text{ for all } j \quad \text{(sign restriction for negative null form)}$$

where μ_j are the multiplier values at the KKT point.

Successful solution of the dual problem assures that these conditions are satisfied for the approximate problem, not for the original problem. Dual optimality gives feasibility of the approximate primal, and recovery of the primal values **x** from the dual solution enforces the approximate primal's optimality. The approximate problem conforms with the original problem only at the point of approximation, so at any point except the true solution of the original problem, satisfaction of the KKT conditions for the approximate problem will not guarantee their satisfaction for the original problem. The main implications here are twofold. First the values of the multipliers are estimates that may be quite wrong if the point is away from the solution. Second, a constraint with positive μ_j at $\mathbf{x}_k$ may not be satisfied as a strict equality, i.e. it is thought of as active but the point does not lie on the constraint surface.

When membership in the extended active set $\mathscr{E}$ is considered, it is reasonable to suggest that constraints which both have positive multipliers and are satisfied with an equality at an acceptable degree should be given priority over those that only have positive multipliers. In other words, the value of $C_{1,i}$ should be influenced by the value of $p_{i,k}$. Alternatively it can be suggested that, in the intermediate iterations, $C_{1,i}$ should be determined by the degree of satisfaction of the transversality conditions rather than the positivity conditions.

The discussion below refers to stress constraints. Other types of constraints may be treated in a similar manner when appropriate.

In a modified strategy several decisions will depend on how far the current constraint value is from its bound, and whether that value is feasible or infeasible. Recall that the set $\mathscr{V}$ includes all constraints that are violated at the preceding iteration, and that the cardinality of $\mathscr{V}$ can be controlled by the value of ε_v. Making ε_v smaller than unity will artificially increase the number of constraints in $\mathscr{V}$. Although this may be desirable in terms of the cardinality of a particular set $\mathscr{E}$, it can be misleading in subsequent interpretation of constraint violations and contradict results based on the value of the feasibility tolerance ε_f. For example, if $\varepsilon_v < 1$ and

$$1+\varepsilon_f > p_{i,k} > \varepsilon_v \tag{33}$$

the constraint is feasible, but it will be regarded as infeasible for the active set strategy. This may be acceptable when this information is used only for determining the cardinality of $\mathscr{E}$ or $\mathscr{A}$. However, in a knowledge-based system that keeps track of each constraint's attributes individually, additional clarification rules will be necessary to resolve this type of conflict. Therefore a clearer approach would be not to allow values of ε_v smaller than 1, and to treat it as a true tolerance, perhaps equal to $1+\varepsilon_f$. In this way the set $\mathscr{V}$ will contain only those constraints truly violated at the given point.

There is merit in wanting to include in $\mathscr{E}$ constraints that appear satisfied, but are close to their bound. These constraints may be assigned an additional bias. Formally,

$$\text{if} \quad 1+\varepsilon_f > p_{i,k} \geq \varepsilon_b, \quad \text{then} \quad B_{i,k} = C_{3,i} \tag{34}$$

Here ε_b is a bound proximity tolerance that has a value less than unity, and $C_{3,i}$ is a priority parameter probably selected so that $C_{1,i} > C_{2,i} > C_{3,i}$. These constraints will belong to a set $\mathscr{B}$ with cardinality $M_{BS}(\mathbf{x}_k)$. This number should be added to $M_{VS}(\mathbf{x}_k)$ in eqn. (16) and other similar estimations of M_{ES}.

In summary, there are three criteria that can be used to include constraints in $\mathscr{E}$ based on local information from the approximate problem, in corresponding order of importance: activity, infeasibility, and feasible proximity to bound. The biased priority index will sort all these constraints both within each criterion and among criteria. At this point the active set strategy has been reduced to generating the set of rules for assigning values to $C_{1,i}$, $C_{2,i}$, and $C_{3,i}$. As mentioned earlier, these numbers are related to the value of $p_{i,k} = (g_i/g_{i,\text{lim}})_k$, so we may let

$$C_{q,k} = c_{q,i} p_{i,k} \qquad q = 1,2,3 \tag{35}$$

and the biased priority index will be given by

$$P_{i,k} = p_{i,k} + B_{i,k} = p_{i,k}(1 + c_{q,i}) \tag{36}$$

where now the $c_{q,i}$ are weight factors associated with each constraint and each inclusion criterion q. The rules for selecting values for these factors can be incorporated in an extension of the PRIMA system.

A note of practical importance for large expensive models should be made at this point. When the counting of constraints is done to determine M_{ES}, constraints may be identified as eligible for inclusion in $\mathscr{E}$ with respect to more than one criterion. Specifically, a constraint may be both violated and have a positive estimate of the multiplier. This double counting increases M_{ES} and additional constraints are included beyond those intended. For many problems this increase will not be significant. In any case, this will be avoided with proper bookkeeping associated with each individual constraint.

In the entire discussion so far the focus has been essentially on how to include constraints in each new $\mathscr{E}$. The entire set is constructed from the start, the only connection with the previous iteration being previous activity based on multiplier estimates. Some attention must be given also to how to exclude or delete constraints from $\mathscr{E}$. The (negative) values of multiplier estimates are used as a source of local knowledge. Further, a good practice is not to add to the active set a constraint which was dropped in the previous one or two iterations. The rationale is that repeated inclusion and exclusion of the same constraint usually leads to cycling and deterioration of convergence properties. Therefore, a competing fourth citerion should be introduced: activity history. Application of this criterion requires also evolutionary knowledge, as opposed to the local and global knowledge of the previous iterations.

Other constraints may be excluded on the basis of global knowledge. A usual example is constraints that describe the limits of theoretical validity of the models used in analysis. Typically, an analysis model is selected or created so that it is sufficiently accurate as a prediction tool for the entire expected range of design variable values. If this range is large, and no other information is available about the vicinity of the optimum, the analyst may be forced to use an elaborate expensive model just for this reason alone. Clearly if the optimization search does not utilize this level of model sophistication, overall efficiency is lowered. An obvious alternative is to use the simpler model but to impose some additional restrictions that limit its range of applicability. This type of constraint must remain feasible and inactive in order to have a meaningful solution. If the iterations continuously find this constraint violated, the initial model may need reassessment

4. CONCLUSION

In the presence of monotonicity, global knowledge about activity may be generated automatically by a program such as PRIMA. The generalized active set strategy described above can be easily coded in a production system environment, and be an extension of or have a link with PRIMA. It is necessary that the system be able to communicate efficiently with the optimizer performing the numerical iterations in FORTRAN, with data passing bidirectionally between the FORTRAN subroutines and the LISP procedures. Current versions of COMMON LISP and OPS83 — the evolution of OPS5 — should make this interfacing less complicated. Although the entire system could be coded in FORTRAN, the opportunity for flexibility would be severely curtailed.

The strategy has clearly many heuristic elements. These may be adapted and augmented on the basis of cumulative experience in a specific knowledge domain, in the spirit suggested by the LAGRANGE program (see Chapter 7). Ultimately, all such systems acknowledge the need to codify and use information that comes from intuition and experience, while pursuing the elusive goal of capturing such information within explicit rigorous mathematical formalisms.

ACKNOWLEDGEMENTS

This research was partially supported by the US National Science Foundation Grant DMC 86-11916. Work on Aircraft Optimal Structural Design was performed at Saab–Scania AB, Aircraft Division, with many ideas motivated by the system OPTSYS and developed with insights by Torsten Bråmå at Saab–Scania. Both support and collaboration are gratefully acknowledged. Excerpts of this work were present at the 1988 GAMM Seminar in Siegen, FRG, and at the 1988 Korea–USA Design Engineering Seminar in Seoul, Korea.

REFERENCES

[1] Arora, J. S. and G. Baenziger, Uses of artificial intelligence in design optimization, *Comput. Methods Appl. Mech. Eng.*, **54**, 303–323, 1986.

[2] Azarm, S. and P. Papalambros, A case for a knowledge-based active set strategy in design optimization, *ASME Paper DE-38*, 1983; *J. Mech., Transm. Autom. Des.*, **106**(1), 77–81, 1984.

[3] Bråmå, T., Weight optimization of aircraft structures: OPTSYS — a system for structural optimization. In Mota Soares, C. A. (ed.), *Computer Aided Optimal Design: Structural and Mechanical Systems*. NATO ASI Series F, Vol. 27, Springer, Berlin, 1987, pp. 971–985.

[4] Bråmå, T. and R. Rosengren, *OPTSYS User Manual, Version 1.2*, Saab Aircraft Division, Saab–Scania, Linköping, November 1987.

[5] Bråmå, T., Implementation of flutter constraints in OPTSYS, *Rep. TKHS-88.82*, Saab Aircraft Division, Saab–Scania, Linköping, 1988.

[6] Brownston, L., Farrell, R., Kant, E. and Martin, N., *Programming Expert Systems in OPS5 — an Introduction to Rule-Based Programming*, Addison-Wesley, Reading, MA, 1985.

[7] Choy, J. K. and Agogino, A. M., SYMON: automated symbolic monotonicity analysis for qualitative design optimization; *Proc. 1986 Int. Conf. on Computer Engineering, Chicago, IL, 1986*, pp. 207–212.

[8] Forgy, C. L., *OPS5 User's Manual*, Department of Computer Science, Carnegie–Mellon University, Pittsburg, PA, 1983.

[9] Li, H. L., Design optimization strategies with global and local knowledge, *PhD Dissertation*, Department of Mechanical Engineering and Applied Mechanics, University of Michigan, Ann Arbor, MI, 1985.

[10] Li, H. L. and Papalambros, P., REDUCE applications in design optimization, *CAD/CAM Robotics and Automation, Int. Conf. Proc., Tucson, AZ, 1985*.

[11] Li, H. L. and Papalambros, P., A production system for use of global optimization knowledge, *J. Mech., Transm. Autom. Des.*, **107**(2), 277–284, 1985.

[12] Li, H. L. and Papalambros, P., An interior linear programming algorithm using local and global knowledge, *J. Mech., Transm. Autom. Des.*, 1988.

[13] Li, H. L. and Papalambros, P., A combined local–global active set

strategy for nonlinear design optimization, *J. Mech., Transm. Autom. Des.*, 1988.
[14] Papalambros, P. Y., Monotonicity analysis in engineering design optimization, *PhD Dissertation*, Design Division, Stanford University, 1979.
[15] Papalambros, P. Y., Knowledge-based systems in optimal design. In Mota Soares, C. A. (ed.), *Computer Aided Optimal Design: Structural and Mechanical Systems*, NATO ASI Series F, Vol. 27, Springer, Berlin, 1987.
[16] Papalambros, P., Codification of semi-heuristic global processing of optimal design models, *Eng. Optim.*, **13**, 235–253, 1988.
[17] Papalambros, P., Enhancements in design optimization problem-solving: a knowledge-based approach, *Rep. TKHS-88,79*, Saab Aircraft Division, Saab–Scania, Linköping, 1988.
[18] Papalambros, P. Y. and Wilde, D. J., *Principles of Optimal Design — Modelling and Computation*, Cambridge University Press, New York, 1988.
[19] Rao, J. R. and Papalambros, P., Implementation of semi-heuristic reasoning for boundedness analysis of design optimization problems. In Rao, S. S. (ed.), *Advances in Design Automation — 1987*, ASME, New York, 1987, pp. 59–66.
[20] Wilde, D. J., The monotonicity table in optimal engineering design, *Eng. Optim.*, **2**, 29–34, 1976.
[21] Wilde, D. J., *Globally Optimal Design*, Wiley–Interscience, New York, 1988.

12

Using artificial intelligence in an open software architecture for modelling in engineering

O. Aunay, S. Aunay, D. Chorlay, G. Touzot and M. Vayssade
Université de Technologie de Compiègne

1. INTRODUCTION

Our aim is to improve the quality of modelling software by facilitating its use in taking knowledge explicitly into account. This chapter proposes to investigate the different possible ways of manipulating knowledge in a modelling software.

For that, we use techniques issued from artificial intelligence (AI). The integration of these techniques can only succeed if the software is built on a suitable architecture.

After reflection on the present state of modelling software, we describe a new software architecture that satisfies the objectives set. We then evoke the problems posed by simultaneous use of AI techniques. We finally present two applications which illustrate the interest of such an approach.

2. PRESENT STATE AND EVOLUTION OF MODELLING SOFTWARE

2.1 Evolution of modelling

Modelling in engineering evolves very quickly for reasons mainly linked to the growth of industrial demand and to the development of computers. We intend first to clarify the main trends of this evolution, emphasizing their impact on the internal organization of modelling software.

2.1.1 *Complexity of physical models*

Physical models in modelling are becoming richer from year to year. In the field of solid mechanics, for instance, software has successively used 'linear elasticity', then 'elastoplasticity', inclusion is the formulation large strains, induced anisotropy, etc. A program should adapt itself to any new physical model by simple replacement of subroutines.

2.1.2 *Coupling*

One can see the appearance of couplings of modellings that were until recently treated separately. For example, simulation of noise generation in an electrical engine couples together models coming from electromagnetism, structural dynamics and acoustics. Software should enable the treatment of coupled problems by simple juxtaposition of the programs treating each decoupled problem.

2.1.3 *Known and unknown information*

New types of modellings modify the traditional separation of information in data and unknowns. This is what is being observed particularly in optimization and in parameter identification of behaviour laws. Software must then manipulate all its information in an unmarked way, without making any distinction, even implicit, between data and unknowns.

2.2 Evolution of modelling software environment

2.2.1 *Variety of users*

Programs are often put in various users' hands, whose objectives and knowledge are different: design technicians, research engineers, researchers, students, developers. Keeping most of its internal modules, the software must be able to adapt easily, by simple replacement of some modules for a particular user's environment. It must be particularly easy to modify the ergonomics of the user interface and the available functions as well as the software flexibility (and no doubt the ease of use, too).

2.2.2 *An algorithm's setting-up language*

Some numerical algorithms are very delicate to set up, particularly those intended for optimization and for the solution of highly non-linear problems in the presence of instability. A good deal of trial and error is often necessary. This setting-up work is expedited by the existence of a modelling language that minimizes the required programming effort. It gives access to all information during the process of calculation, in reading as well as in writing mode.

2.2.3 *Software life-cycle*

Experience shows that the lifetime of much modelling software lies between 10 and 20 years. Over such a period of time, it is very difficult to predict the evolution of the software application fields as well as that of numerical and computing techniques. The necessary modification of the software must be possible in a continuous way, avoiding the appearance of successive releases too different from one another and of imperfect compatibility.

2.2.4 *Co-development*

The development of substantial items of software involves large teams that are often difficult to bring together in the same place. Moreover, one notices that the global efficiency of a development team generally decreases with its growing size after a certain point. The organization of software must facilitate the sharing of the development effort between several distinct teams. This implies that the information required for participation in the development should be as limited and well defined as possible.

2.3 Evolution of computer techniques
2.3.1 Equipment
After early computers adapted to batch processing and those designed for time sharing, one can see networks of heterogeneous calculators becoming widespread. These can rely at the same time on personal computers, workstations, and vector and/or parallel computers specialized in intensive computation. Operating systems offer increasing possibilities for simultaneous use of several processors.

New specialized equipment, such as highly parallel calculators, database processors or pattern recognition units, will soon be added to existing computers. In order to benefit from these new computer systems, modelling software must be able to divide up the operations to be performed on several processing units, using the specificities of each of them at their best. This can be done for example by making several slightly different versions of the same software communicate, these versions running in parallel on various processors.

2.3.2 Documentation, learning and ergonomics
The volume of documentation describing modelling software as well as the learning time necessary to use the software correctly are increasing too quickly. This evolution constitutes one of the major obstacles that limits the diffusion of this type of software. A good internal organization of such software must lead to a documentation as limited and structured as possible, and whose volume increases slowly as the software possibilities become wider. Only autodocumentation mechanisms can guarantee the consistency between a program and its documentation.

The experience gained with new microcomputers and workstations proves that good use of windows, menus and on-line documentation considerably reduces the need for written documentation and the learning time for inexperienced users. However, one notices that experienced users prefer to work with a command language than with menus. General software must then allow the simultaneous use of these two types of user interface.

2.4 Manipulation of knowledge
2.4.1 Transfer of knowledge
A very diverse knowledge is manipulated by modelling software, and its users and developers. It concerns at the same time the fields of application, the numerical methods used, the computer science and the particularities of the specific software. Mechanisms of creation and transfer of knowledge

between these three actors are complex, sometimes explicit but most often implicit. Let us summarize some of these mechanisms:

— When writing software, the developer has in mind a field of application and several hypotheses which are more or less explicitly defined. He partially transfers his knowledge into the user's documentation. This is reflected in the programming and in the test runs.
— The choices made by the developer concerning the software's internal structure are partially described in its internal documentation which is intended for subsequent developers. Only the software itself contains in its coding lines all the information concerning it, but in a form difficult to deal with.
— The software user needs very diverse knowledge to use it efficiently. This comes at the same time from his experience in the field of application, from his previous experiences, from the software's external documentation, and from his ideas about the software's capacities, limitations and drawbacks.
— Each software execution not only generates results but also knowledge concerning, for instance, the results' accuracy and reliability, and the algorithms' numerical difficulties. This knowledge is perceived in a raw form by the user who partly utilizes it subsequently and passes it on (sometimes) to the developer; the latter possibly modifies the documentation and the software accordingly.
— Finally, let us recall that any modelling requires a certain *a priori* knowledge of the physical system under study. This materializes by simplifying hypotheses, mesh choices, etc.

2.4.2 Modelling software and knowledge

In present modelling software, knowledge is not manipulated as such. It appears in a diffuse way at every level of the elaboration and use of software. To simplify the task of users and developers, it is desirable to explicate the mechanisms of creation, storage and use of knowledge as well as to provide specific tools for knowledge manipulation such as:

— insertion in the software of the knowledge accumulated by its user or explicitly given by its developers;
— automatic learning by exploitation of results from modelling runs.

If, furthermore, the software has some knowledge about itself, it can help its users in several ways:

— By answering questions such as the following: what can I do now, and why?; if I modify one item of information, what does it imply?; how do I reach a database state or how do I build up an item of information?; if there are several paths to reach one given state, which one shall I pick up, under which criterion, and why?
— By protecting the user against illegal or dangerous operations.
— By helping each user to build his personal view of the software adapted to its specific needs. The software should offer controllable transparency

between complete opaqueness (the user does not know anything and the software is leading him), and total transparency (the user knows everything and the software allows him any manipulation, including its own modification).

2.5 Conclusion
There are two kinds of problems that can be inferred from previous sections which complicate the use and development of modelling software. They are related,

— one to the architecture of software, and
— the other one to the way software handles knowledge.

In what follows are described the choices we made in these two fields: software architecture and knowledge management.

3. PROPOSED ARCHITECTURE

3.1 Preliminaries
Every modelling process is divided into a set of elementary operations clearly identified and defined (called 'commands' in the following). These are chosen so that combinations of elementary operations enable modellings as varied as possible to be performed. The software is organized in such a way that:

— each elementary operation can be activated by the user, thanks to an appropriate input command language;
— every command can be executed at any time; it produces a valid result as long as its inputs are available and valid.

To achieve this result, a certain number of constraints must be respected, in paticular:

— each command must clearly know its input and output data;
— the software must make a clear distinction between the executable code and the data it is working on;
— a command must be able to execute another one.

This organization then offers a lot of advantages:

— The software description consists mainly in the description of its commands.
— Any command size can be used from the simple $I=I+1$ instruction to the complete resolution of a modelling problem. Only simplicity of use and efficiency constraints limit this flexibility.
— The software evolution is achieved by adding, suppressing or replacing commands; it can therefore be very smooth.
— The set of commands constitutes an extensible and evolutive modelling language.

3.2 General organization

The general software organization that we are putting forward involves three large functional blocks as illustrated in Fig. 12.1:

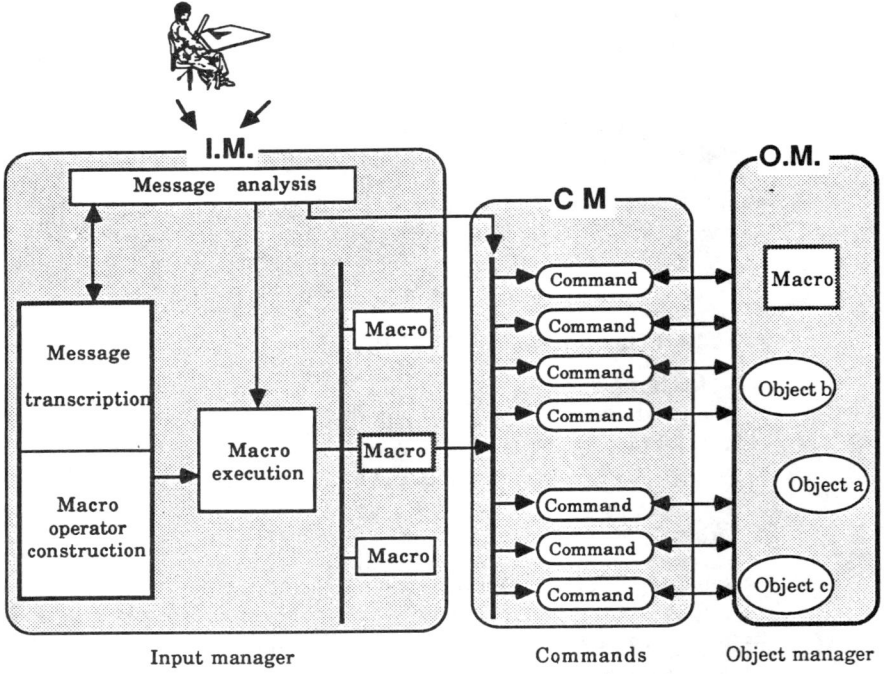

Fig. 12.1 — General software organization.

— The input manager (IM) interprets the information provided by the user from different devices (keyboard, mouse, menus, vocal input, etc.). This leads to the execution of one or several elementary commands with the passage of the required arguments.
— A set of elementary commands (CM) and macro commands execute all the operations needful for any available modelling: calculations, information and knowledge management, display requests, etc.
— The object manager (OM) handles all the software information gathered together in object form.

The three functional blocks communicate through two interfaces (IM–CM and CM–OM) which in fact constitute the only intangible norms of the software. Each block (IM, CM, OM) can evolve independently as long as it respects interface norms.

3.3 The object manager
3.3.1 Memory management
The object manager is in charge of storing and retrieving any information manipulated by the software. This information can be both largely varied (nodes, elements, logical links, production rules, etc.) and very bulky (large matrices, numerous calculation results). Even on computers with a large memory, management must be space saving. In particular, it is essential to allocate space dynamically during the work process and to recover the space which is no longer in use. The management of this unique workspace used by every part of the software constitutes the first function of the OM.

3.3.2 Object and identifier handling
The objects are created, destroyed and moved by the OM in the workspace. An object puts together every item of information logically related to an entity such as a node, an element or a matrix and whose access at the same time is desirable. An object is given a certain type (node, element, real matrix, etc.) that enables the OM (and thus the developer and the user) to interpret the information inside the object. The second function of the OM is therefore the manipulation of such typed objects. The objects are placed sequentially in the workspace at creation time:

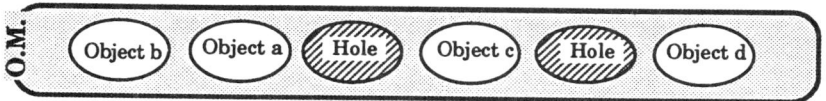

The holes left by the destruction of objects can be put together so as to regenerate one or more large contiguous free spaces; this is done by a more-or-less complex strategy based on the displacement of objects. After compacting:

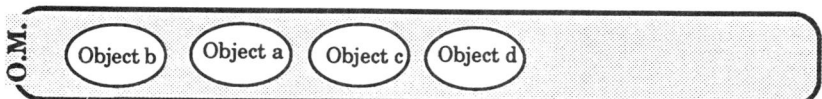

This object mobility makes their position (address) in the workspace useless for designating or retrieving them. That is why the OM assigns each object a unique identifier (Id) at creation time that will stay unchanged during the object's life. The OM maintains the correspondence between the address (variable with time) and the identifier of each object thanks to a particular object: the address table (Ad.T).

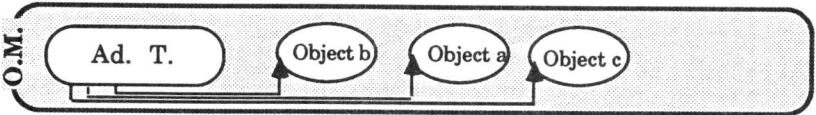

The identifiers are used for the creation of logical links between objects and for all mechanisms of access to objects. They introduce, in fact, a decoupling between memory management (address) and access mechanisms (identifier).

3.3.3 Objects internal strcuture
Each object is made up of a descriptor and a set of attributes and zones whose number and significance depend on the object's type. The descriptor is managed by the OM and contains in particular the identifier, type and length of the object. An attribute may be an integer or real number, a character string or an identifier of another object. Zones are similar but their length is dynamically variable; they are automatically managed by the OM. The content of attributes and zones is used and modified by the different commands of the software. Here are a few examples of attributes and zones of 'node-type' objects:

— an 'integer' attribute: the external number of the node;
— a 'real' attribute: the x-coordinate of the node;
— a 'reals' zone: the values of all the node's degrees of freedom;
— a zone containing the identifiers of elements connected to this node (the length of this zone remaining zero as long as the inverse connectivity has not been built).

For each known object type, one disposes of a description of each of its attributes and zones including in particular the following:

— the type of information (integer, real, character, identifier);
— the initial values and lengths (for zones);
— a name facilitating the attribute's designation by the user;
— the offset between the attribute and the object's origin.

This parameterized definition of the object's internal structure enables easy evolution of this structure with time, without changing already-written programs. It has the advantage of giving very rapid access to the attributes of an object as soon as its identifier is known. The OM offers the necessary tools to create new types of objects dynamically.

3.3.4 Object access mechanisms
The different software commands need to access objects, that is, to find their identifiers (thus their addresses). Certain commands (assembly of the stiffness matrices for example) need to access objects frequently and thus have to optimize speed. Others (display of the content of an object) give greater importance to the friendliness of the access mechanism: the user wants to manipulate the objects with the name he has chosen. The OM therefore offers two major access mechanisms to objects:

(1) An extremely rapid mechanism (cost equivalent to one or two indirect addressings) is meant to access data required by operations located in the internal loops of computation. This is based on a very flexible organization of objects in networks, each object being able to point

towards one or several other objects through attributes or zones containing identifiers.

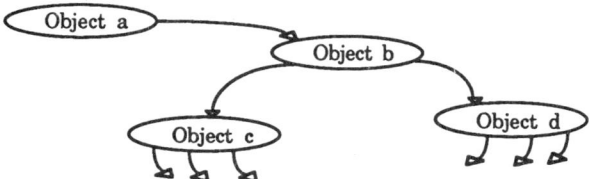

This mechanism is used in particular to establish logical links between elements, nodes, materials properties, boundary conditions, etc. Different data structures (lists, trees, index, stacks, etc.) can be built up from such a network-type structure. The OM offers construction and updating tools for such data structures (this is particularly useful for artificial intelligence or for the representation of parameterized complex geometrical shapes).
(2) A mechanism of friendly access to objects is provided that uses names that user or programmer may assign to objects. In this case, a particular object and a hashing-type technique enable the relation between name and identifier to be established.

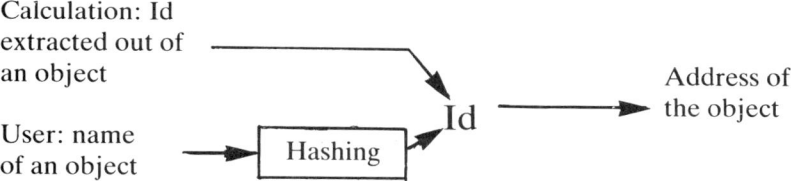

3.3.5 Evolution
In short, the OM constitutes a library of subroutines that offers the following services, while respecting the constraints of efficiency required by large modellings:

— it manages a memory space which contains typed objects of dynamically variable lengths;
— it creates and uses object patterns that enables the interpretation of the content of each type of object;
— it offers a rich set of access mechanisms to objects, which enables the building up of 'multilayered databases' that manipulate, in several ways, the same set of objects.

The marked modularity of the OM makes it a tool with a high evolution potential: it is easy to add new access mechanisms to objects, to create complex data structures (particularly useful for artificial intelligence), to add distributed database functionalities (parallelization), etc.

4. USE OF ARTIFICIAL INTELLIGENCE TECHNIQUES

4.1 · Organizational problems

The use of a new computer techniques born from research on AI in modelling software developed in procedural languages such as FORTRAN or ADA causes technical as well as conceptual problems. We are presenting here the major choices we made in this field.

4.1.1 Choice of application field

Among all the modelling operations in which one desires to manage knowledge explicitly, those where AI techniques can immediately give some help must be chosen: it is not a question of carrying out research in AI but a question of using results acquired in this field in order to improve the modelling efficiency.

A simple choice consists in looking for operations complex enough to justify the use of AI but which only require well-known and well-mastered present techniques. Among all the computer novelties born from AI research (languages, inference engines, frames, semantic networks, object-oriented programming, etc.) only a few have reached a truly industrial efficiency (security of use, ease of development, execution speed, reasonable memory space, etc.). Care must be taken not to use artificial intelligence techniques to solve problems that are within the capacity of good classical programming.

Among the fields where AI is useful, one can give the following:

— aid in modelling control;
— taking into account the user's experience in application fields in which it plays an important role;
— control of sensitive algorithms;
— taking into account external criteria such as production technological rules and production costs.

4.1.2 Choice of interaction level between modelling software and artificial intelligence

One of the difficulties consists in defining the relative positioning of AI and 'classic' modules (calculation, geometry, graphics, etc.), as well as their possible interactions. At this level, one can distinguish two possible implementations:

— The AI part constitutes the main system frame and uses modelling subroutines as 'black boxes' called up in the process of reasoning. In other words, AI controls the software.
— The AI part is made up of 'advising modules' capable of solving logical-type problems. For that, one uses knowledge about the formulated problem and possibly about the software itself; therefore one is able to give an answer to the user as well as to an internal modelling module.

Our point of view is that the second method enables a better interaction between AI and modelling. In this method one considers AI modules as a library of additional tools, similar to a graphical mathematical library of additional tools, similar to graphical or mathematical library. Moreover, this choice enables an optional use of AI modules depending on the problem difficulty or the user's desires. Finally, its modular aspect fits in well with the proposed architecture and gives greater flexibility in the control and development of the different software parts.

Here we want to be allowed a 'sociological' comment: the first method corresponds to an 'imperialistic' approach of AI. It is often the method spontaneously considered by AI researchers. It is the logical consequence of AI's novelty as an industrial tool and of the lack of support AI specialists still suffer from. It is difficult to make them understand the interest of the second method because it seems to them that it considers AI as an auxiliary technique. Conversely, modelling researchers do not really admit to be completely under the control of AI. The second method discussed above balances the relations between traditional modelling techniques and AI techniques.

4.1.3 *Information storing: the databases problem*

The modelling modules, as we saw in the second part, manipulate data structures based on objects stored in a database. The AI modules also need a database to store their knowledge and intermediary results. However, the same 'database' expression designates totally different computer implementations. Effectively, information storage related to AI uses particular techniques such as knowledge bases, fact bases, frames and objects (as an object-oriented programming), whereas modelling modules have more classical needs.

Therefore, one is *a priori* faced with two types of databases working differently but containing complementary or redundant information about the same problem. Moreover, it is important to note that these two databases logically constitute only one: the AI modules use information from the 'modelling database' and the modelling modules exploit results coming from the 'AI database'.

From that, one can derive two ways to approach the problem: either building up only one integrated database containing both modelling and AI information or keeping two separate databases specialized in the management of information of different types. Let us examine these two approaches:

— *An integrated base*: the available data structures must be efficient for both modelling and AI. This implies specialized access mechanisms for AI. This method offers a certain consistency in the way that it centralizes information and standardizes its management. If one uses a specialized AI language, there are two ways to achieve this goal: to manage all data by this language or by the language used in modelling modules. This

leads to incompatibilities in terms of efficiency because of external management of data.
— *Two separate bases*: the consistency problem between the two bases must be solved. Effectively, depending on activated modules (modelling or AI), one of the two bases is not updated. Therefore special mechanisms must be developed to make a database evolve depending on modifications that occur in the other. This complicates the management of databases because part of the processing must be 'duplicated'. Thus, here too, one may fear a certain loss of efficiency.

Depending on the relative volume of AI and modelling databases and the quantity of calculations to be carried out by the two types, these two options may be more or less efficient. Our experience in this area is not sufficient today to give a well-argued answer. Nevertheless, this is an implementation and performance problem that does not inhibit research, as long as the databases' consistency and integrity are guaranteed.

4.1.4 Choice of a conceptual model to store knowledge

Knowledge used by an AI module inside modelling software can be of two kinds: external to the software (production process, available computer resources, work context, etc.) or internal, that is concerning the operation of the software itself.

Certain AI modules are meant to reason about the present and desired state of the modelling database (the goal); they must then possess a model of possible evolution of this base. This model must include, obviously, the architecture of the modelling software and its operation.

In the first studies so far conducted, we have chosen as the software logical model a graph joining entities (broad notion of an object present in the database) and operators (broad notation of a modelling operation, for example a command). The entities are the nodes of the graph, the oriented arcs joining the nodes represent the operators. The different states of the database, corresponding to the state of progress of the modelling work, are characterized by the existence of entities constituting the graph. Solving a problem then boils down to the route of a path on this graph.

One can adopt different 'granularities' of entities and operators for the graph, depending on the case being processed. This, pushed to the limit of its logic, brings a fundamental novelty in the use of modelling software: the possibility of putting forward a hypothesis and simulating its logical consequences. This simulation is the work of AI modules but does not lead to any actual modelling action before confirmation by the user.

4.2 Technical problems

As we just saw, the use in the software of AI and classical modelling modules brings a certain number of structural problems. More, if one uses a specialized language (LISP, PROLOG) to develop AI modules, other technical problems may appear.

4.2.1 Choice of AI language

To write AI modules, the developer has two possibilities: either choosing a tool specialized for AI (language or expert system generator) or keeping the language used for modelling (most often FORTRAN) so as to use only one computer language. Let us examine the advantages and drawbacks of each of these possibilities.

— *A specialized tool*: this solution is certainly the most natural for developing an expert system. It also guarantees a good performance at the AI level. Effectively, the developer can find every tool he may need in the market of languages or of expert system generators: front, back or double chaining, object-oriented programming, simple or multiple inheritance mechanisms, knowledge bases developing tools, etc. On the other hand, this solution inevitably leads to technical difficulties of 'interfacing' the AI tool and FORTRAN, as well as portability difficulties.

— *FORTRAN*: the second solution consists in developing one's own AI tool in FORTRAN language. This eliminates the previous computer problems since all the software is written in one language. However, the writing of an inference engine, or of any other AI tool, is a complex task, reserved to specialists. Moreover, it does not correspond at all with our approach which consists in using existing AI tools and not in developing new ones.

The choice between these two possibilities depends, of course, on the context and the development team's competence. We opted for the first solution by developing our AI modules in PROLOG and writing the necessary interfaces with FORTRAN and the OM.

Let us note that, in the architecture we are proposing, the different commands are independent of the data. This leaves complete freedom to the developers in the choice of the programming language; the only constraint lies in the possibilities of calling on the OM from the considered language.

4.2.2 Integration of 'modelling' codes and AI codes

This choice being made, we are left with the problem of overcoming a certain number of technical difficulties. The first one consists in making both programming languages communicate. We want to be able to call on a PROLOG predicate from a FORTRAN subroutine and vice versa. For that, we have chosen a PROLOG opened towards other languages. This is the 'Prolog de Delphia' which is written in C and which is thus perfectly interfaced with this language. To make FORTRAN and PROLOG communicate, we use an intermediary layer of C functions that enable cross-calls and ensure the passage of arguments. The application portability is thus that of the three FORTRAN, C and PROLOG compilers.

4.2.3 Information sharing

One uses then two systems each of which possesses its own mode of information storage (totally incompatible). To simplify, we call 'database' the storage tool associated with the modelling modules and 'facts base' the tool associated with the PROLOG modules. The whole of the information of these two bases must be accessible from every module of the software, whatever its nature (Fig. 12.2).

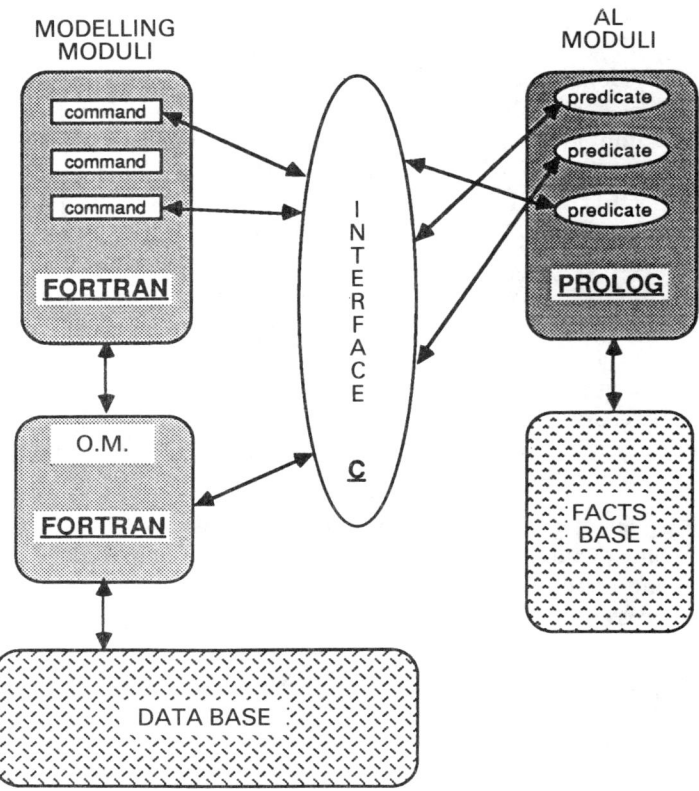

Fig. 12.2 — Integration of the AI modules.

The following four situations are possible:

— A FORTRAN subroutine wants to access information in the database: no problem, it uses the object manager directly.
— A PROLOG predicate wants to access information in the facts base: no problem, this access is directly managed by PROLOG.
— A FOTRAN subroutine wants to access information in the facts base: it calls a PROLOG predicate (through the C interface) which accesses the information and returns it to the calling subroutine.
— A PROLOG predicate wants to access information in the database: it

calls a FORTRAN subroutine (through the C interface) which accesses the information and returns it to the calling predicate.

In this way, the problem of information sharing is partially solved by the passage of arguments in the reciprocal calls between FORTRAN and PROLOG.

However, one difficulty remains inasmuch as both languages manipulate data of very different natures, respectively numerical and symbolic. Thus, it is not enough to transfer the information, it might need to be adapted to the language. This problem essentially arises when a PROLOG predicate uses a fact (symbolic or logical information which can take two values: right or wrong) evaluated from the database. In this case, the predicate must obtain this numerical information and apply to it an algorithm or a heuristic enabling it to build up the fact it needs. This solution may look a little complicated, but the only alternative would be to duplicate every item of information in each of the bases, which would strongly penalize the application.

4.3 An example of artificial intelligence use for the assisted control of modelling software

The optimal use of a design system is related to knowledge exploitation and to the user's experience. The quality of the results obtained with a modelling system is thus partly dependent on the user's competence level, that is, external to the system itself. This is a very important aspect which is generally neglected. Therefore can one come up with the following question: is it possible to reintegrate part of this knowledge into the system by using artificial intelligence tools?

4.3.1 Required knowledge for modelling software use

The knowledge manipulated is bulky; it concerns at the same time the software application field (mechanics, fluids, etc), geometry, calculation methods, computing (hardware or software), work context, commands, etc.

The nature of this knowledge is diverse. Theoretical and bookish knowledge as well as more intuitive knowledge coming from the user's experience are used.

One can clasify this knowledge according to its origin and abstraction level. At the most general level is the physical problem we intend to trreat. From there, we establish a mathematical model (equations) which is itself transcribed into a numerical model. Finally, at the lowest level of this hierarchy, is the modelling software. All this knowledge may well come up at any time in the user's reasoning. Thus a user continuously transcribes information related to this problem seen from several points of view, in order to reason at different levels.

The manipulated facts are also distributed and of different natures. The database associated with the design software contains numerical data, some of which describe explicit information that may be directly transcribed in facts form for an AI system. In other cases, information may be present in an implicit or diffuse way. This kind of information, often obvious for the user (visualization, commonsense) is much more difficult for an AI system to access. Finally, certain kinds of information are not memorized at the software level but are only present in the user's mind. To make up for the absence of a facts base, in the classical sense of the term, one uses information present in the database and then transforms it into symbolic data. The database evolves with time; objects are created, suppressed, modified. However, the existence of an object in the database is not enough to guarantee the validity or utility of the information it carries. This is the responsibility of AI modules.

In a modelling system, one desires to introduce knowledge about its own operation such that it becomes capable of estimating the process (series of commands) that enables it to reach a goal fixed by the user. There is no question of suggesting, in an authoritarian way, the user's choices, but rather the user is discharged of certain tasks. The sequencing of operations is not specified any more by the user but is suggested by the software itself. The latter has effectively a certain 'initiative' which enlarges its scope of use. These new possibilities of dialogue imply the use of some kind of free interface (natural language, dynamical menus, etc.) exceeding the simple set of commands.

4.3.2 A conceptual model of the modelling system

To reach these objectives, a conceptual model of the modelling process must be defined. This model concerns the process in its entirety and not only the software which is just the computer implantation of it. In other words, knowledge from fields such as geometry, mechanics, thermics, physics, calculation methods and computing has to be taken into account despite of their entire or partial absence from the software.

As we have already seen, one describes the modelling process in terms of two types of objects: the operators and the entities. This information constitutes a virtual graph, although it is not actually stored in this form. Practically, one builds up every operator–entity relation in the knowledge base.

An operator corresponds to any modelling operation, for example defining a boundary conditioning, assembling the stiffness matrix, visualizing the lines of isostresses, performing a non-linear complete computation.

In the same way, an entity corresponds to any information manipulated by an operator such as the following facts: the forces vector is assembled, no elementary property is defined, the computer used is a workstation of low performance, etc.

The software commands are particular operators and the objects present in the database are particular entities. However, one can notice in the example above that the notions of operator and entity are deliberately very general and exceed the simple description of the software. One can also use logical operators to wwhich is not associated any action. We describe two operators of this type: 'imply' and 'constitute'.

The 'imply' operator joins entities in the general sense of implication.

For example, the relation 'the calculus is non-linear of the material is plastic' will be stored in the following way:

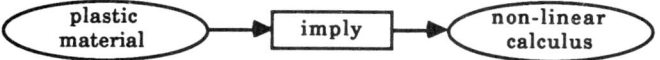

The operator 'constitute' enables the description of complex entities to be refined. Thus the description 'the external environment of the field under study include the external forces and the boundary conditions' will be stored in the following way:

The use of these logical operators thus enables one to code in the description graph of the modelling system knowledge rules of the same type as that classically used in expert systems.

An operator is an oriented link between entitites. However, one can run along this link in both directions, and thus obtain the 'dual' (or reciprocal) operator without having to define it explicitly.

4.3.3 Setting up the control module

Before detailing the setting-up of this control module, let us examine its principal functions:

— The detection of every event that occurs at the level of the software whatever its origin (commands, menus, natural language, vocal commands, etc.). This function ensures the updating of the 'facts base', which is required for the software to be controlled.
— The transcription of events into an 'internal language' understandable by control module.
— The construction of a macro operator enabling a given goal to be reached.

The detection of an event is performed by the input manager.

The transcription is ensured by a set of specialized interfaces, based on a knowledge of the input means used, as well as on a description of the modelling process itself. The main tool, called the control engine, is able to build up the sequence of operators (macro operator) enabling a given goal to be reached. To enable that to be done, it is based on the conceptual model of the modelling system that we have already described. It is important to notice that a link must exist between the process description of the modelling used at the time of the transcription of an event and that used by the control module. This is essential to interpret an event and then to make it 'understood' to the control engine (Fig. 12.3).

The building up of a macro operator depends on the existence of the

244 DESIGN OPTIMIZATION [Pt. II

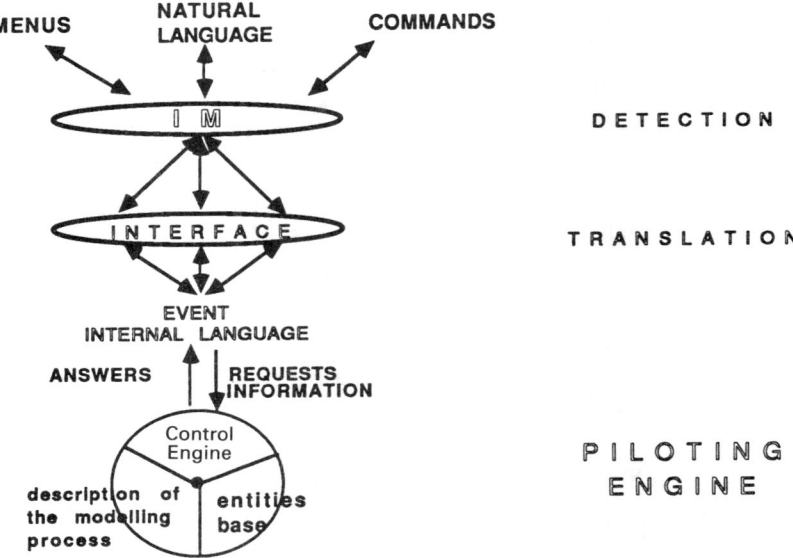

Fig. 12.3 — Control module.

entities it manipulates. One may consider that the entities occurring in the system description are a photograph of the state of advancement of the modelling process. The description 'entities base' is used for the set of entities existing at a given time; it is equivalent to a 'facts base'.

To simplify, one may say that the entities–operators graph describes what it is theoretically possible to do, and that the entities base constitutes an instantiation of this graph, corresponding to the modelling steps actually performed.

Such a system modifies the relationship between the user and the software.

First of all, the user expresses himself in his own language; it is no longer necessary to know the commands exhaustively. The software possesses 'intelligent behaviour' or a certain 'initiative'. It becomes possible to detect certain inconsistencies in the user's modelling process, which offers, among other possibilities, an obvious educational aspect. Another very interesting perspective consists in managing simultaneously several entities bases. One of them is related to the database, and the others correspond to hypotheses proposed by the user and investigated by the system (Fig. 12.4).

It is therefore possible, in this case, to propose hypotheses and to see immediately what the consequences of them would be. The choices of the user concerning the material physical properties, the geometry, the solicitations, the meshing, etc., are then made under better conditions since they come after a verification of the hypothesis' validity. It is then truly a matter of computer-aided design.

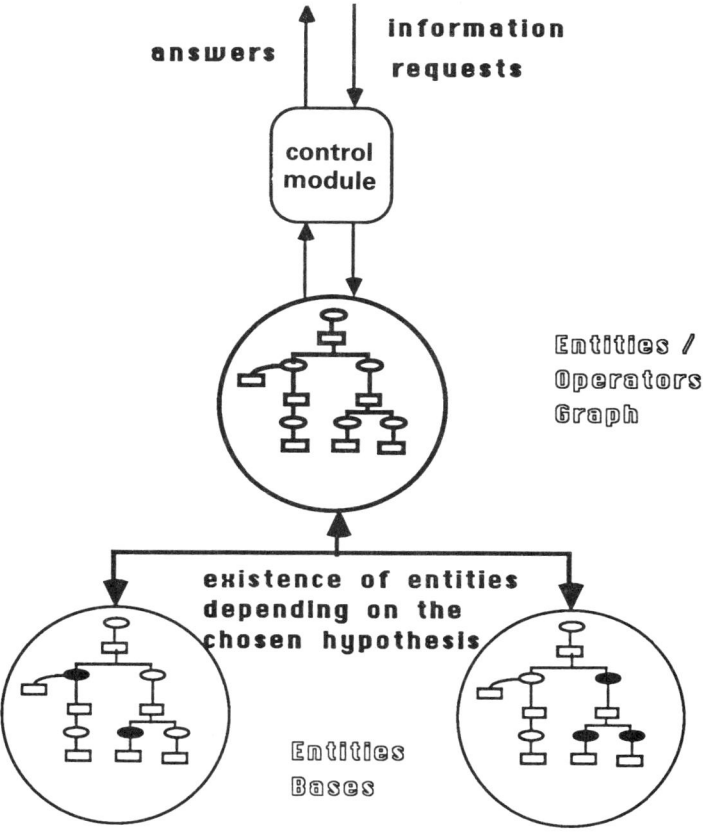

Fig. 12.4 — Exploration and comparison of hypotheses in the entities–operators graph.

4.4 An example of artificial intelligence use to account for technological constraints

4.4.1 Description of the problem

The application described in this part concerns the optimization of casting pieces designed for cars, essentially of crankcases, suspension arms and exhaust collectors. The casting pieces are drawn according to certain criteria and a very accurate schedule. This schedule groups together a set of constraints with respect to the piece's service behaviour. Certain constraints concern more particularly the mechanical resistance, others the problems of location inside the vehicle, and others the production costs.

To verify the mechanical behaviour of the piece, it is modelled and then the behaviour is calculated. This modelling represents the mechanical problem to solve. The designer proposes a shape, on which calcualtions are performed. The calculation results indicate good or bad behaviour of the proposed form (Fig. 12.5).

From these results, optimization may have two objectives:

— to minimize the total mass of the piece, with limits according to the von Mises criterion;
— to minimize the mean are maximum value of the von Mises criterion, the total mass having been fixed.

In this example, the considered structures are of a thin shell type, whose mean surface is fixed. The values of the thickness around this mean surface are taken as optimization variables [1].

First of all, the considered problem is one of mass optimization. Its objective is to minimize the mass, satisfying the mechanical constraints imposed on the piece by limits on the von Mises criterion.

In the proposed architecture, additional optimization commands have been written. The software enables at the moment either the mass, or the stresses to be optimized. This optimization modifies the piece's shape, by making the wall's thickness vary.

4.4.2 Technological constraints

The technological constraints do not depend on the function of the piece itself but on its production technique. They take the form of production constraints, which concern the know-how of the foundry technique, and the cast needed to produce the pieces.

The foundry enables highly standardized pieces of complex shapes to be produced at low cost. The productivity criteria increase the technical constraints imposed by the casting process. To produce a foundry piece, it is necessary to make a mould that will give it its external shape. The mould is made up of two parts: two subframes, filled up with sand, in which one prints the external form of the piece. This is achieved by using two model plates (MP), which are the exact representation of the external shapes of both sides of the piece. The joining surface between the two subframes when they are shut is called the plane of joint (which is not necessarily made up of only one plane, but often of several planes).

The internal shape of the piece is defined by one or more nuclei. The nucleus is also in sand (so that the inside of the piece can be emptied once it is made) and is placed in the mould before it is shut. It is made of a box with two parts, comparable to the foundry mould; it is the nucleus box (NB). The two parts of this box define exactly the internal shape of the piece to be cast. Sand is injected then solidified in the nucleus box. The box is then opened and the nucleus extracted and then placed in the mould of the piece to be cast.

The first production constraints that the caster needs to respect, so as to guarantee the internal and external shpes of the piece, consist in being able

— to extract the nucleus out of its production box,
— to remove from the mould the model plate from the print.

For this, one introduces tapers (NB and MP), that will guarantee the extraction of the nucleus and of the MPs from their moulds. It is in fact a

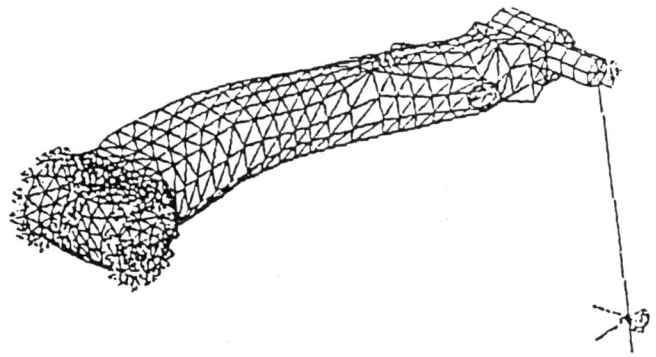

Fig. 12.5 — Mesh, loading and boundary conditions.

matter of giving a slight angle (a few degrees) to the vertical walls relative to the plane of joint. This production constraint is absolutely essential, because it conditions the production of nuclei and prints. It must be completely checked by the pattern of the foundry piece (Fig. 12.6).

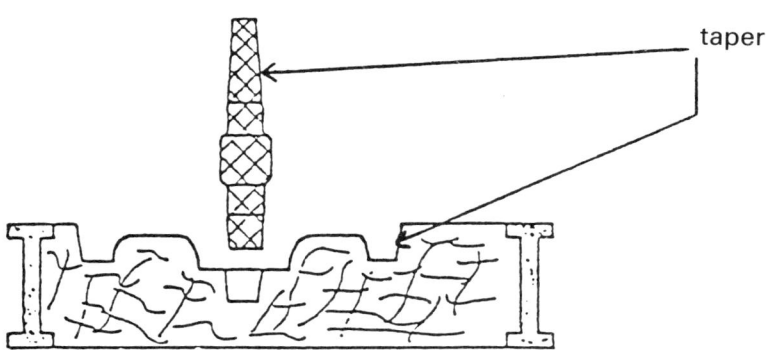

Fig. 12.6 — Tapers.

The second constraint concerns shrink holes. This is clearly a more delicate phenomenon in the foundry process. It arises from the fact that the metal shrinks when solidifying. Poor metal feeding in the zones during the process of solidification causes gaps (from a few microns to a few millimetres) in the material, these are called shrink holes. They generally happen in the parts where there are large variations of thickness.

The solution to this problem consists in feeding correctly the piece of metal by placing, close to important volumes, a metal and heat reserve. This reserve is called a feeder. Several feeders are generally required to cast a piece correctly. Once the piece is cooled down, the feeders are cut off, since they are not part of the shape of the final piece (Fig. 12.7).

Fig. 12.7 — Feeders.

It is obvious that feeders are expensive, because they are molten metal volumes which are only used to feed the piece during its cooling down. The number of feeders is therefore limited by the caster, who must strictly guarantee that a piece without any shrink hole or micro-shrink hole will be obtained.

It is therefore necessary to check that every part of the piece is well fed, and that no shrink hole can appear. This is one of the rules that a caster follows when drawing a piece: if an overthickness is inevitable, a feeder is then necessary.

To the technological constraints of the foundry know-how, constraints of production costs are added. Effectively, certain foundry techniques cannot be put into practice in highly standardized production, because of their high cost. Thus, the set of constraints defines a technically and financially acceptable domain, that the drawing must respect. For that, one changes the financial constraints into technical constraints by limiting the production processes used to those that are the easiest and least costly.

The analysis of the production constraints for tapers and shrink holes leads to the formulation of rules. For example, with the shrink holes, the relation existing between the value of an overthickness, its area and its distance from a feeder conditions the appearance of a shrink hole, and might be transcribed into a set of rules. A rule may be expressed in the following form:

— If the distance to the feeder is large, and if the overthickness is average, and if the area is large, then the risk of a shrink hole is large.
— If the distance to the feeder is small, the risk of a shrink hole is nil.
— If the distance to the feeder is average, and if the overthickness is small, and if the area is large, then check the risk on neighbouring elements.

4.4.3 AI modules and optimization process interaction

The first objective of optimization, which consists in minimizing the mass of the piece, is not modified, the goal always being to make the volume of melting necessary for the production as low as possible. On the other hand, technical constraints on production are added to the constraints on mechanical behaviour.

The relations between the objective of optimization and the constraints to respect is thus modified, inasmuch as the solution domain is restricted by production technical constraints.

It is difficult to express the knowledge about the foundry in a procedural language. It is certainly possible to code a rule such as that given previously in FORTRAN, but it would be difficult to manipulate, to adjust and to improve. Moreover, this rule is part od a non-deterministic set of rules, which represents the necessary knowledge for the detection of a shrink hole risk. The number of rules amplifies greatly the problem of coding into a procedural language. It therefore seems to us more interesting to model this knowledge in an AI-type language such as PROLOG.

The constraints of the optimization process thus split into two categories: the constraints of mechanical behaviour, taken into account in the optimization process, and the technological constraints, which are checked by an AI module.

The verification of technological constraints must be frequent, so, as to avoid the convergence of the optimization process to a solution not satisfying the production technical possibilities. Control should be carried out every time a thickness allocation is proposed by the optimization process. This process is iterative. and converges to the solution by proposing, at each iteration, a thickness allocation closer to the optimum than the previous one. The control module must therefore be resorted to inside the optimization iteration, immediately after the thickness allocation calculation (Fig. 12.8).

The rule's execution first requires the building up of the necessary facts. For that, the module is going to take information from the modellization database. The user can be asked for other facts, notably the facts which only concern the casting problem and which cannot be deduced from the modellization data. These facts are prepared in a separate module before the optimization system is started. The exploration of the rules is then launched to determine the risks of violation of the different technological constraints. Correcting rules can also be set up to correct the risks previously deduced. The resulting facts are then interpreted to be transcribed in actions on the optimization variables, or on the cosntraints applied to the optimization process.

4.4.4 *Computer implementation of the AI modules*
Each module set up in the prototype is given a particular task. The size of the modules remains reasonable, and enables a high execution speed. The data for the modellization type are searched for through the intermediary of the object manager (OM). The AI specific data are memorized under fact form by the system appropriate for the language used.

The preparation and verification modules are programmed in the 'Prolog de Delphia' language. Some reasonings have been modelled with the expert

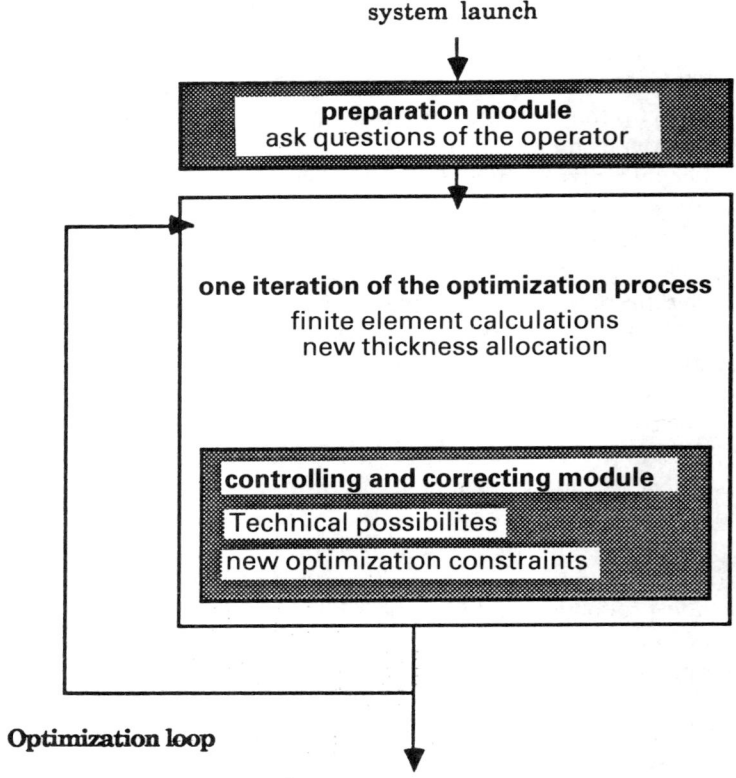

Fig. 12.8 — Organizational diagram.

system generator Mylog, from the same company. It uses the 'Prolog de Delphia' environment. The interfacing between FORTRAN and PROLOG has been described in section 4.2.3.

The advantages of the solution used are interface speed and efficiency: it avoids duplication of data, and allows access to modellization information is rapidly from an AI module as from a FORTRAN subroutine.

On the other hand, using two very different languages in the programming method effects the production of pertinent facts for the foundry problem. Effectively, every item of information of the object manager is organized with modellization in mind, and it is sometimes very difficult to rebuild relatively abstract facts.

The user can be asked for certain facts, because they are invariable during the whole optimization process: the position of the feeders, of the plane of joint, etc. On the other hand, other facts require an approach of the pattern recognition type; this is the case for overthickness detection. This has to be executed for each new overthickness allocation, and usually uses a process of the caster's global vision on a drawing. It is very difficult to imitate this process with rules; pertinent criteria must therefore be found to determine the necessary facts.

The contribution of advising modules in this type of optimization is obvious: it enables knowledge which is difficult to represent in the form of a program written in a classical language to be modelled. The interfacing and cohabitation of modules written in different languages do not pose an insoluble technical problem. Nevertheless, the big difference of data types makes transcription problems appear. Equally, the power of AI-type languages is sometimes insufficient when it is a matter of solving problems requiring pattern recognition processes.

5. CONCLUSION

Wee have presented a new architecture of modelling software, respecting a few very simple basic principles and satisfying the objectives mentioned at the beginning of this chapter.

We have shown how appropriate data management enables easy and efficient insertion of AI-type modules. This integration has been illustrated by the setting-up of AI modules inside an optimization loop.

The simplicity of the software's internal organization makes it efficient, easy to deal with and rational to use. It enables the elaboration of a conceptual model of the software, allowing knowledge describing its use to be taken into account explicitly.

The control of sensitive algorithms, the automatic synthesis of macro commands, the intelligent assistance and the parallization of automatic calculations are as many evolutions, made possible by the exploitation of this knowledge.

REFERENCES

[1] Beldi, M., Optimisation de la loi d'épaisseur des structures minces, *Thèse à paraitre*, UTC, 1988.

[2] Favard, R. and Marx, G., Vers une architecture de système de CAO intégrant un langage d'intelligence artificielle: Prolog, *Actes de la 7ème Conference Européene sur la CFAO et l'infographie*, MICAD, Tome 1, Hermes, 1988.

[3] Felippa, C. A., Database management in scientific computing. I. General description, *Comput. Struct.*, **10** 1979.

[4] Felippa, C. A., Database management in scientific computing. II. Data structures and program architecture, *Comput. Struct.*, **12** 1980.

[5] Jacobsen, K. P., Fully integrated superelements: a database approach to finite lement analysis, *Comput. Struct.* **16** (1–4), 1983.

[6] Law, K. H., A parallel finite element solution method, *Comput. Struct.*, **23** 1986.

[7] Laporte, F., Sur la conception d'un systeme expert d'aide à l'utilisation de logiciels scientifiques, *Huitème Colloque International sur les*

Méthodes de Calcul Scientifique et Technique, Tome 2, INRIA, December 1987.
[8] Rajan S. D. and Bhati, M. A., Data management in FEM-based optimization software, *Comput. Struct.* **16** (1–4), 1983.
[9] Sadar Amin Saleh, H., Artificial intelligence and computer aided design in civil engineering. Sriram D. and Adey, R. (eds) *Applications of Artificial Intelligence in Engineering Problems*, Vol. 2, Springer, 1986.
[10] Sayettat, C., Systemes experts et aides à la conception, *Calcul des Structures et Intelligence Artificielle*, PLURALIS, 1987.
[11] Skreekanta Murty, T. and Arora, J. S., A survey of database management in engineering, *Adv. Eng. Softw.*, **7** (3), 1985.
[12] Skreekanta Murty, T., Shyy, Y. K. and Arora, J. S., MIDAS: management of information for design and aanlysis of systems, *Adv. Eng. Softw.*, **8** (3), 1986.
[13] Sriram, D., Maher, M. L. and Fenves, S. J., Knowledge-based systems in structural design, *Comput. Struct.*, **20** 1985.
[14] Taig, I. C., Expert systems and finite element analysis, *Finite Elem. News*, **4** 1986.
[15] Taig, I. C., Expert aids to finite element system applications. In Sriram, D. and Adey, R. (eds), *Applications of Artificial Intelligence in Engineering Problems*, Vol. 2, Springer, 1986.
[16] Taverniere, P., Marcovitch, J., Knopf-Lenoir, C. and Aunay, O., A propos de l'utilisation de l'intelligencce artificielle appliqueé à l'optimisation de formes sous contraintes techologiques, *Calcul des Structures et Intelligence Artificielle*, PLURALIS, 1987.
[17] Touzot, G., Réflexions sur l'architecture des logiciels de modélisation, *Calcul. des Structures et Intelligence Artificielle*, PLURALIS, 1987.
[18] Trousse, B., Bénéfices d'une approche orientée objet pur un environnement de CAO, *Actes de la t^{eme} Conférence Européenne sur la CFAO et l'infographie*, MICAD, Tome 1, Hermes, 1988.
[19] Wilson, K. G. and Rogers, T. R., A proposal to the NSF to establish the center for theory and simulation in science and engineering, *Internal Rep. CTSSE 84–2*, Cornell University, September 1984.

Index

active set strategy, 210
advice system, 37
AI, 16, 37, 59, 121, 227, 236
aircraft structure, 209
Airy stress function, 128
atom, 78

backward inference, 61, 77
BB framework, 78
BBL language, 78, 90
bifurcation point, 107
blackboard system, 75

chunk, 100
cluster, 100
clustering techniques, 151
CNF (conjunctive normal form), 84
combinatorial explosion, 77
computational mechanistics, 121
CONMOD, 75
constitutive modelling, 75
constraints, 137, 147, 184, 209

dam stability, 67
damage assessment, 70
deep knowledge, 60
deflection constraints, 137, 219
demon, 80
dependability, 39
design, 15, 159, 184, 209
discrimination algorithm, 46
dynamic analysis, 114

EAL, 182
ESD (event sequence diagram), 41
evidential support, 50
EXADS, 182
expert system development tool, 61
explanation facility, 103, 154
EXSYS, 159

finite element method, 97, 181
flutter constraints, 137, 220
forward inference, 61, 77
frequency constraints, 219
fuzzy set, 44

GENIUS, 98

hazard prevention, 65
hierarchical modelling, 49
Horn clause, 51
hypertext, 88

IDEAS, 182
inference networks, 23, 188
INSIGHT, 123
interface between program parts, 64, 104, 232, 240
interval probability, 42

KBES (knowledge based expert system), 16, 37, 64, 75, 97, 122, 159, 181
KKT (Karush-Kuhn-Tucker) conditions, 222
knowledge base, 20, 52, 81, 187
knowledge classification, 48, 60
KS (knowledge source), 78, 91
KSAR (knowledge source activation record), 79

LAGRANGE, 135, 224
limit point, 107

machine learning, 45, 147, 158
macro-command, 104
MACSYMA, 126
mathematical programming, 135, 147, 184, 209
measure of the truth, 43

membership function, 44
monotonicity principle, 213
multistory buildings, 159

NASTRAN, 138
Nemark method, 114
Newton-Raphson procedure, 108

objective function, 136, 147, 184, 209
open software architecture, 227
OPSYN, 182
OPS5, 212
OPS83, 224
optimization, 135, 147, 184, 209

PCFEAP, 98
plane stress isotropic thermoelasticity, 93
POF (pattern-object form), 83, 94
prefix predicate calculus notation, 83
PRIMA, 212
probability, 42, 165

QSEIS, 15
quantitative-qualitative transformer, 110

reasoning mechanism, 17

resolution inference, 77, 85
risk control, 37
rules, 22, 56, 61, 100, 123, 143, 165, 189

SACON, 16
safety assessment of concrete dams, 65
seismic behaviour, 15
semantic net, 100
shallow knowledge, 60
signal analysis, 54
snap-through, 108
solid mechanics, 75
stability in elastostatics, 107
statistical machine learning, 147
stress constraints, 137, 217
structural system, 17, 59, 69, 135, 181, 209
structures subjected to explosions, 69
SUSY language, 145
symbolic manipulation, 129
symbolic-numeric environment, 59

tangential stiffness matrix, 108

uncertainty, 42, 101, 200
user interface, 26, 135, 244

Venn diagram, 44

MAR 1 3 1991